Taurus

Aldebaran

Hyades

The first book to blend scientific observation of the Sun, Moon, stars, and planets with self-improvement and mindful reflection

Thousands of years ago, human storytelling, planning, navigating, and timekeeping often centered on the stars. Yet today, in our fast-paced modern world, we have lost this deep connection to the cosmos that was once central to our daily lives.

Offering a unique combination of science and self-reflection, *Astronomical Mindfulness* reconnects us to space once more, guiding readers through the fundamental ways our planet moves through the universe and how these motions determine our perception of time and place. You'll learn methods of observing the celestial sphere and how to contextualize your place in it—deepening your connection to the Earth and helping you become more informed, engaged, and mindful every day.

You don't need to climb a mountain, visit an observatory, or even own a telescope. From an apartment rooftop to a city park, from your backyard to the window by your desk, the open skies are accessible to everyone. *Astronomical Mindfulness* is an essential tool in our modern world, empowering us to be more present and more relaxed simply by looking up toward the stars.

ASTRONOMICAL MINDFULNESS

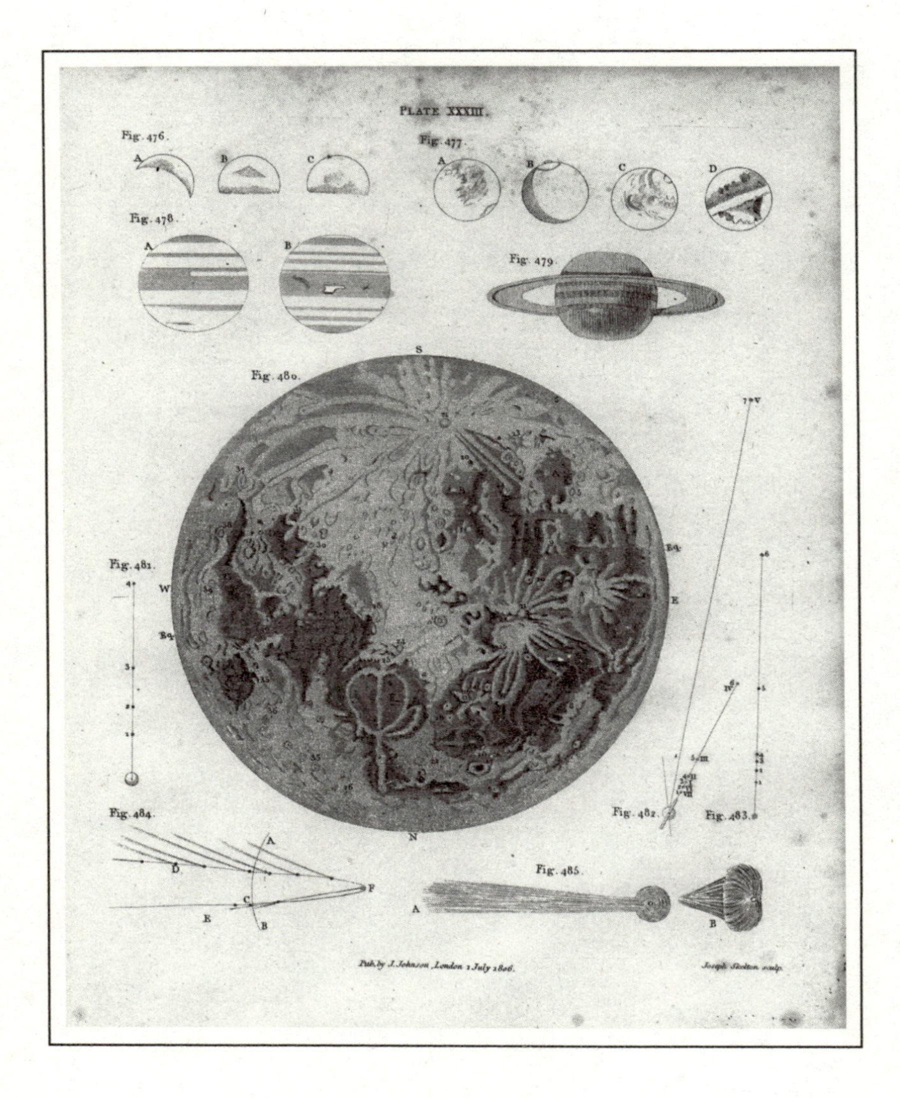
PLATE XXXIII.
Fig. 476.
Fig. 477.
Fig. 478.
Fig. 479.
Fig. 480.
Fig. 481.
Fig. 482.
Fig. 483.
Fig. 484.
Fig. 485.
Pub. by J. Johnson, London 1 July 1806.

ASTRONOMICAL MINDFULNESS

YOUR COSMIC GUIDE TO RECONNECTING WITH THE SUN, MOON, STARS, AND PLANETS

CHRISTOPHER G. DE PREE AND SARAH SCOLES

HarperOne
An Imprint of HarperCollins*Publishers*

HarperCollins books may be purchased for educational, business, or sales promotional use. For information, please email the Special Markets Department at SPsales@harpercollins.com.

FIRST EDITION

Unless otherwise noted, all illustrations by Claire De Pree
Illustration on page ii by Sergey Kamshylin/AdobeStock

Library of Congress Cataloging-in-Publication Data has been applied for.

ISBN 978-0-06-304132-5
ISBN 978-0-06-323935-7 (ANZ)

22 23 24 25 26 LSC 10 9 8 7 6 5 4 3 2 1

To my parents, Gordon and Lenore De Pree,
who inspired me to be a writer, and showed me how to
be mindful of each moment.

—CHRISTOPHER G. DE PREE

To my family, who told me to keep looking up.

—SARAH SCOLES

CONTENTS

INTRODUCTION

Back when humans were living in communal caves and tribal encampments, we told stories about the stars, even though we didn't know what they were. When we started sailing, we used these same pinpricks of light to estimate our own location. When we began planting, we counted on the Sun's position and the appearance and disappearance of certain constellations to remember when to plant and harvest our crops.

Now we tell stories about ourselves, and we use advanced tools to navigate. For the most part, someone else does the planting for us, and seasonal changes are marked by back-to-school sales and spring-cleanings. While we (arguably) understand ourselves and the planet better than we did long ago, we've mostly lost our connection to the cosmos, and to our planet. We may have apps that tell us when we can see the next Full Moon, but how often do we actually go look at it?

Many would say that to look up at the sky is a modern luxury, not a necessity. Worse, it sounds like a waste of time in a productivity-driven culture. As a result, the press of daily life has largely erased our understanding of the motions of the Sun, Moon, stars, and planets. We are go, go, go, but where, exactly, are we going? Where have we come from? What does it all mean? Attention to the cosmos just might provide some answers.

Plenty of astronomy books can tell you about the constellations and the positions of the planets, but unless you're taking a college course or you're just especially curious, there isn't a pressing need to explore the galaxy. However, there are benefits that have nothing to do with the reasons our ancient ancestors looked to the skies: looking at the night sky brings you into the here and now. When you're focusing on the stars, you're not scrolling through social media, zoning out in front of the TV, or staring at your out-of-control inbox. Everything slows down. You breathe easier. You feel calmer. This book is a practical guide to understanding the fundamental ways in which our home planet moves through the solar system, and how these motions determine our perception of time and space. But perhaps more important, it is a book that will help you be more mindful about your presence in the universe.

"Mindfulness" has become a buzzword, often clouded by bland, corporate jargon, or overcomplicated and lumped in with the practice of meditation, which can feel daunting and unattainable. But the true definition of mindfulness is actually quite simple. Thich Nhat Hanh, often called the "father of mindfulness," put it this way in *The Miracle of Mindfulness*: "Don't drink your tea like someone who gulps down a cup of coffee during a work break. Drink your tea slowly and reverently, as if it is the axis on which the whole Earth revolves—slowly, evenly, without rushing toward the future. Live the actual moment. Only this actual moment is life."[1] In other words, it's really focusing on where and how you—corporeally and cognitively—are situated in and interacting with your surroundings, noticing yourself, the environment, and the relationship between the two. Often mindfulness brings to mind terrestrial examples—like that half-full glass of water that you're sipping. But mindfulness can apply equally to celestial matters. Where and how are your brain and body relating to the particular happenings in the sky above? How do the photons of light warming your skin actually connect you in a real way to the Sun's roiling surface?

The acts of looking up, noting what's going on, and contemplating how it all relates to you are mindful by their very nature, because mindfulness—by *its* very nature—is about being open to, curious about, and attentive to what is happening around and above you right now. When you are mindful in your observations, you are living life in the moment. And understanding and noticing the sky above can help you to be more informed, engaged, and mindful during each day and night.

These actions are particularly important in a world that is increasingly fast and loud, and in a society that tells us our value lies in what we produce and the status we can achieve. In their book *Mindfulness*, Mark Williams and Danny Penman suggest that instead of eliminating pleasurable activities to make more time for productivity and work, we should intentionally find ways to make our daily lives more enjoyable. They propose three such ways: (1) do something pleasurable, (2) do something that will give you a sense of satisfaction or mastery over your life, and (3) act mindfully.[2] Astronomical observations can accomplish all of the above. Simply gazing at the clear night sky will provide a sense of pleasure. Following through on any exercise in this book will provide a sense of mastery and achievement. And careful observation of your present place in this moment is one way to act mindfully and be a more-aware citizen of the universe.

The takeaways in this book are profound and yet primitive, just like mindfulness itself. In some ways, mindfulness is just a fancy term for "paying attention." At its core, it is both simple and significant. "*Mindfulness* is synonymous with awareness," writes expert Jon Kabat-Zinn. "We are colossally out of shape when it comes to perception and awareness, whether oriented outwardly or inwardly or both. We get back in shape by exercising our faculties for paying attention over and over again, just like a muscle."[3] This book can help you to be aware of your interaction with the universe through observation, to exercise those muscles day and night, and to feel more alive not just as a person on Earth but as a human in the universe. "Astronomical mindfulness" is simply practicing mindfulness when you observe the universe, in the same way that you might mindfully make tea or wash the dishes. When you're observing the universe, that should be what you're doing. When you're looking at the Moon, say to yourself, *I am looking at the Moon*, even though it seems obvious. Be present in that magnificent moment.

HOW TO USE THIS BOOK

We don't really have to notice anything about the sky to lead a successful modern life, other than perhaps menacing clouds or impending darkness. And because we don't have to, and we're busy, we usually don't. But the practices in the following pages will provide a reason to pay attention anyway, and to let your thoughts flow through, accepting them, as you might in traditional meditation.

The book is organized into five parts, one that introduces you to your surroundings—"Getting Oriented"—and then four more that focus on observations related to the Sun, Moon, stars, and planets. You don't need to work through each part in order, but it's probably a good idea to start with the "Getting Oriented" chapters. Each chapter is rated with one, two, or three asterisks. These reflect the order of complexity and time required, with one asterisk being the least complex and taking the least time to complete. Each activity also has a mindfulness exercise, with one or two of the following labels to signify its focus:

PRESENT MOMENT exercises help you focus on what's going on in your body or in your environment right now.

CLOSE OBSERVATION exercises ask you to notice the details of something you might otherwise let slip by.

BREATHING exercises help you bring attention to the present and to your circumstances through that most human of tasks: respiration.

PERSPECTIVE exercises give you the space to contemplate your place in the cosmos.

VISUALIZATION exercises help you create a mental scene that situates you in the universe.

At the end of each part are two "interlude" chapters, one that explores a particular local astronomical tradition and one that presents a profile of an expert in that tradition. The work of these Native and Indigenous astronomers and scholars of history incorporates a more mindful relationship to the sky than modern professional astronomy typically does. In many ancient traditions, astronomical knowledge is both practical and spiritual. Bridging these different traditions, and knitting them together, can make knowledge of the sky more meaningful, not just more abundant.

Finally, the appendices in the back of the book dive deeper into particular practical topics—like why stars have hard-to-pronounce names or appear to be different colors, and what apps you might want to download to further your observation and learning.

COSMICALLY CONNECTED

Taking stock of where planets show up relative to one another and to constellations, and how that changes as the seasons and years pass, brings home the idea that all these worlds, as well as the other objects we can see in the sky, are connected to one another, and to us, even in the face of vast distances. Beyond that, now more than ever, it's important to invest in our connection to the world around us, and to explore what connects us versus what divides us. Because we're all residents of the same universe, at the same moment in time. And we are all inconceivably tiny in the grand scheme of things.

So now that you know where we're headed, walk outside, take a deep breath, look up, and start to notice the sky above you. "While washing the dishes one should only be washing the dishes," said Thich Nhat Hanh. "The fact that I am standing there and washing these bowls is a wondrous reality. I'm . . . conscious of my presence, and conscious of my thoughts and actions."[4]

Let's bring that same awareness to the sky. It's time to reconnect with the Sun, Moon, stars, and planets, with our past, and with our own minuscule and massive place in the universe.

I. GETTING ORIENTED

IN WHICH YOU BECOME ACQUAINTED WITH YOUR LOCAL AND UNIVERSAL SURROUNDINGS

How closely have you observed your surroundings and the parts of the sky that are visible from where you live? The practices in this section of the book will help you become acquainted with your local surroundings, with direction, and with the structure of the sky. These fundamental and introductory ideas lay the groundwork for the observations and reflections that follow.

REFLECTION ON TECHNOLOGY

If you pull out your phone when you're supposed to have your sights set on Venus, that has to take away from your experience of Venus, right? As noted mindfulness expert Jon Kabat-Zinn puts it, "With digital devices and smart phones, we are now able to be so connected that we can be in touch with anyone and everyone at any time, get texts and calls, or check email anywhere and at any time."[1]

But what about mindful use of technology? Is that possible? It sounds counterintuitive, but the devices haunting our purses and pockets can actually help us to be more oriented in time and space, and able to engage in open-minded awareness of the heavens above us.

In all our efforts to be mindful, it makes sense to limit the beeping, pinging, and serotonin-inducing interruptions. Many technology experts suggest limiting notifications so days are not constantly peppered with the notice that someone liked our tweet, Facebook post, Instagram story, TikTok video, or whatever we'll be putting online in fifteen, twenty, or fifty years.

But smartphones and smart watches allow more than fleeting validation: they also let us access basically all human knowledge—including what's up with the current sky. While there are thousands of night-sky apps available, three general categories are extremely useful:

- compass apps,
- apps showing sunrise/sunset time and lunar phase, and
- constellation mapper apps.

See appendix B for some specific suggestions (page 193).

While the gradual increase in knowledge that comes from looking at the night sky is perhaps the best long-term approach, these apps let newbies quickly identify constellations and asterisms (groups of stars), and make it easier for them to become more proficient in identifying celestial objects.

YOUR PERSONAL OBSERVATORY

Your first reaction to the title of this exercise might be "What observatory? I don't have an observatory! Aren't observatories staffed by scientists, located at 14,000 feet, and operated by robotic computer systems?"

And, sure, most are like that.

But the point of this exercise—and, in a way, the point of this whole book—is to demonstrate that your observatory is wherever you are. Your eyes can be powerful tools to understand the night sky. Your hands can measure distances. And long before anyone built a telescope, many millennia's worth of humans—huddled around campfires at night, with zero smartphones—had a deep and abiding understanding of the motions of the Sun, Moon, stars, and planets.

You might live among skyscrapers or in a remote farmhouse or in a suburban neighborhood. You might have a big open field near you or a rooftop area on your apartment building, humming with AC units. You might have a porch that faces west or east. Or your house might be ringed by trees, with just a circular peek at the sky above you.

Wherever you live, whatever your observing conditions, you can become more familiar with the night sky and its majestic, regular patterns and motions. And in becoming more aware of your environment, and the activity above you, you can connect with both your local surroundings and the objects that are unimaginably far away. So let's get started.

WHAT YOU'LL NEED

A daytime or nighttime sky
This book, a notebook, or a sketchbook
A pencil or pen

WHAT YOU'LL DO

In this exercise, you'll pick a spot near home that you can return to often, noticing its pros and cons as an ad hoc observatory. You'll make a quick sketch of the horizon around you. While you might not be able to do all the exercises in this book from this precise spot, you should be able to do many of them here, and finding a place close to where you live is a good way to get yourself outside more often, just like picking a gym along your commute route means you're more likely to go to that spin class.

THE DETAILS

Do this whenever the mood strikes, day or night.

- Find a location near where you live, perhaps in your front or backyard, in a park across the street, on the sidewalk in front of your apartment, or on a patio or balcony with an open view of the sky. The important thing is that your location be easy to return to, so that you're less likely to say, "I'd rather Netflix and chill."

- You may know them already, but if not, use a map (online or old-school) to figure out the rough directions of north, south, east, and west at your location. They don't have to be precise—just close!

- From where you're standing (or sitting) in this location, make a quick sketch of the horizon in the four cardinal directions (north, south, east, west). If, say, north is blocked by a big Walmart, that's okay. Just go ahead and indicate what you see in your sketch.

- Now look directly overhead. What do you see? Is it daytime or nighttime? If you stepped outside at night, can you see any stars? If you went out during the day, what does the sky look like? Is it clear? What color is it? Any cumulonimbus clouds (the thunderheads that rise high into the sky) floating around?

- Check back on the sketches you have made here and decide whether your observatory is a good vantage point for all the coming exercises. Some exercises, for example, require you to be able to see the western horizon. Some require you to find the Moon.

- Use the following space provided to describe and draw what you see around you and in the sky above.

KEEPING IT SHORT

You can do this exercise even more quickly if you simply know where north, south, east, and west are at your location. Just note which of these directions give you a clear view of the sky!

THINGS TO PONDER

- Have you ever looked at the night sky from where you live? What have you noticed?

- Do you think people who live in large cities notice the night sky less? What might explain that?

- Where have you lived or visited that gave you the best view of the nighttime sky? What was it about that place that made a difference?

- How does thinking of your home as an observatory change your perception of where you are?

- Did you look at the night sky more when you were a child? Do you remember what you thought about it then versus what you think about it now?

MINDFULNESS EXERCISE (PRESENT MOMENT)

Find a comfortable place to sit outside, somewhere that you might make observations from eventually. Close your eyes and try to focus on the present moment. Be aware of the sounds you hear, the sensations in your body, the smells around you. The present is fleeting, and always continues to pass by, but it is also all that you have. Focus on this moment in time in all its detail for a minute or so before you open your eyes.

DIRECTIONS

We've all been there: You wake up alone in the woods in the middle of the night. Where are you? What direction is north? How did you get here? And how can you find your way home?

Well, maybe you haven't been in this particular situation. But orienting yourself is essential in connecting with your local environment. Sometimes it's easy. The roads of many cities are laid out in a clear, compass-like grid. The numbered streets in downtown Chicago, for example, run roughly east–west, and the broad avenues go approximately north–south. But what if there are no streets? What other clues can you use to point yourself in the right direction?

WHAT YOU'LL NEED

A daytime or nighttime sky
This book, a notebook, or a sketchbook
A pencil or pen
Your Star Wheel (see appendix A, page 187) to recognize some major, easy-to-identify constellations if you're doing this exercise under a nighttime sky

THE DETAILS

For the daytime observer:

- Look for the Sun in the sky. If it's just come up, you're peering approximately east. If it's about to set, then you're gazing west. Congratulations! You know directions.

- If you'd rather go out in the middle of the day, and you live north of the equator, you'll find the Sun traveling along an arc in the southern sky. Congratulations again: you know where south is. Spin halfway around and you'll nab another prize: you know where north is.

- You can also get some info from the ground-based world. Look for moss on any trees around you. Mosses like to live in the shadows, so in the Northern Hemisphere, they grow on the northern sides of trees, where they throw the most shade. Voilà: north!

- On a sunny day, go out into a park or wherever you're aware of the rough directions. In the following space, make a drawing of any clues that help you find directions.

For the nighttime observer:

- On a clear night, go outside and try to find the Big Dipper in the northern sky. You should be able to catch it most times of the year, unless you live close to the equator or in the Southern Hemisphere. It may help to use your Star Wheel.

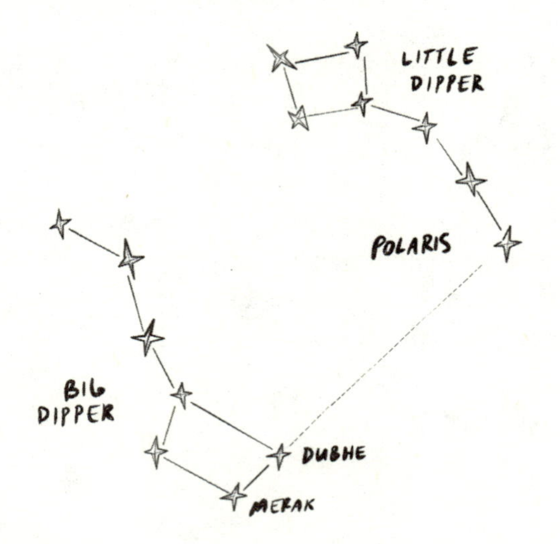

The "pointer stars" Merak and Dubhe can be used to find the North Star, Polaris.

- The "pointer stars" at the end of the Big Dipper's "bowl" point to the North Star, Polaris (though note that Polaris is visible only in the Northern Hemisphere). The diagram above shows how to find the North Star once you've got the Big Dipper.

- The North Star, like its name implies, tells you where north is. And it also tells you where on the globe you are. If you were stand-

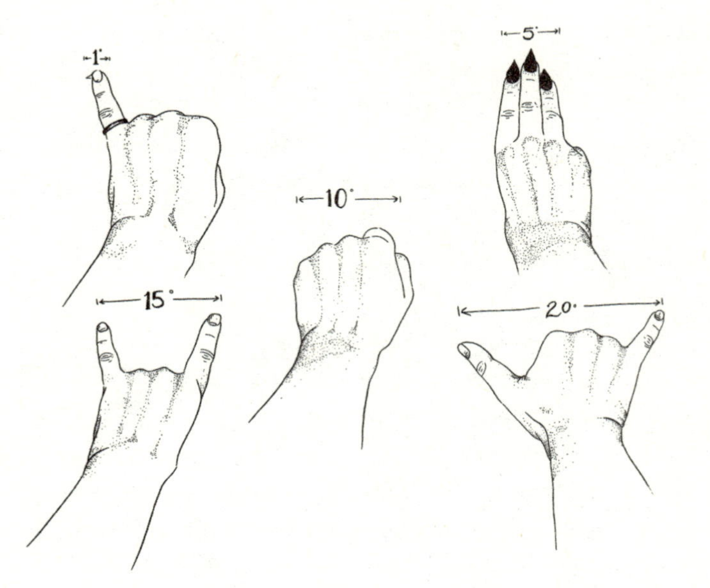

Your hand (extended at arm's length) can be used to measure a variety of angles (shown) on the sky.

ing at the North Pole, Polaris would be directly overhead. If you were standing on the equator, Polaris would be just on the horizon. Chances are, you're somewhere between those two extremes. Once you've found the North Star, stretch one arm out in front of you. Use your hand and fingers to measure the number of degrees straight up from the ground (facing north) to the North Star. That number is your latitude, or how far you are, in degrees, from the Earth's equator. The drawing above, in particular the part showing that your fist measures 10 degrees, can help you to measure the angle with your hand.

- If you want, look up your actual latitude and compare it to the one you calculated yourself.

THINGS TO PONDER

- How could knowing directions help you to survive in the wilderness?

- When might it be useful for you to know your latitude, and how might it have helped ancient navigators?

- How do you think sailors figured out their latitude when they were far south of the equator—sailing around the southern tip of Africa, for example—and they could not see Polaris?

- Does it change your relationship to directions when you're able to orient yourself without opening a map on your phone?

MINDFULNESS EXERCISE (CLOSE OBSERVATION)

The next time you're on a walk in the woods in the Northern Hemisphere, take a look at the trees around you. Notice the bark and any mosses sprouting on the northern, generally shady side of the trees (in the Southern Hemisphere, mosses favor the southern side of the trees). If you're not able to find any moss, take a moment to focus on the tree's details. Make a mental note of any unusual colors, textures, or smells.

WHERE IN THE SKY ARE THE PLANETS?

One of the many joys in getting to know the nighttime sky is being able to identify familiar groups of stars. Picking out these shapes—like the curl of Scorpius's tail or the dash of Orion's belt—connects us not only to the cosmos but also to thousands of years of human history. Generation after generation, people have looked up at these dot-to-dot pictures, which look essentially the same as they did millennia ago. Different societies may have had different names for what we now generally call Orion, for example, but people knew that distinctive collection of stars—what we now see as a belt and sword—thousands of years in the past.

There have always been "wanderers" among the stars. Before the advent of the telescope, these bright points of light—planets, it turns out—looked just like the stars, except they moved across the sky, always hugging the same path the Sun takes. The first telescopes (over four hundred years ago) magnified these objects enough for observers to see them for what they are: distant worlds, very much like our own, yet also strikingly different.

Many ancient societies spent time and effort tracking the motion of the planets relative to those stars, discovering, for example, that Jupiter traces a circle in the sky about every twelve years.

In this exercise, you'll learn where to find the planets and follow their journeys across the sky, like human observers have done for millennia.

WHAT YOU'LL NEED

A clear nighttime sky
This book, a notebook, or a sketchbook
A pencil or pen
A smartphone (optional)

WHAT YOU'LL DO

You'll find the general path that the Sun takes through the stars during the year. Scientists call this path the "ecliptic." You'll also find out whether you can see any planets from where you are and make note of how they shift over time. The rate the planets move, and your ability to track them, will depend on which planets you observe and where they happen to be in their journey around the Sun. A smartphone app that allows you to locate some planets is helpful for this exercise but is not necessary.

THE DETAILS

- If you live in the Northern Hemisphere, then the Sun meanders through the southern sky. If you live in the Southern Hemisphere, it stays in the northern sky.

- Make a drawing of your local sky in the direction of either north or south, depending on your hemisphere. Pick a spot where you can observe the sky over a period of time.

⊙ Draw the horizon from east to west, making note of buildings, trees, etc. It doesn't have to be a Rembrandt—a cartoony sketch will do! It might look something like this:

If you're using a smartphone app:

- A variety of smartphone apps allow you to basically view the night sky with labels. One of the simplest is called Star Rover. Opening this app in the default mode will give you a view of the sky, with planets and constellations superimposed.
- Direct your phone to the southern sky (if you live in the Northern Hemisphere). See where the planets are, and then go outside and check them out with your own eyes!

If you're not using a smartphone app:

- Look to the south (at night) and see if you can locate any particularly bright "stars." Those might be planets!
- If you don't see any bright beacons that might be planets, or you're unsure, search the internet for "planet rise and set times." This will give you a list of times of day when planets will come into view and go away. For example, from my location today, Venus will set at 22:01 (or 10:01 p.m.). So if I go outside tonight at perhaps 8 p.m., I should have a nice view of Venus in the western sky and know sort of where the ecliptic is!

- Using the information from your smartphone app, or from your observations of the night sky, draw the location of as many of the planets as you can see.
- Make note of the planets' positions among the stars and constellations.

- Once you've found a planet or two, it should be easy to find them at the same time the next night. They won't have moved much! So in the previous example (not using a smartphone), I could go out for several nights in a row at the same time (about 8 p.m.), when and where I know Venus will be visible, and notice how Venus moves a little in the sky.

- As you observe and make note of the planets over multiple nights, you should be able to draw the ecliptic in the southern sky: just sketch, on your original drawing of your local sky, an arc that roughly connects all of the planets you've been able to find.

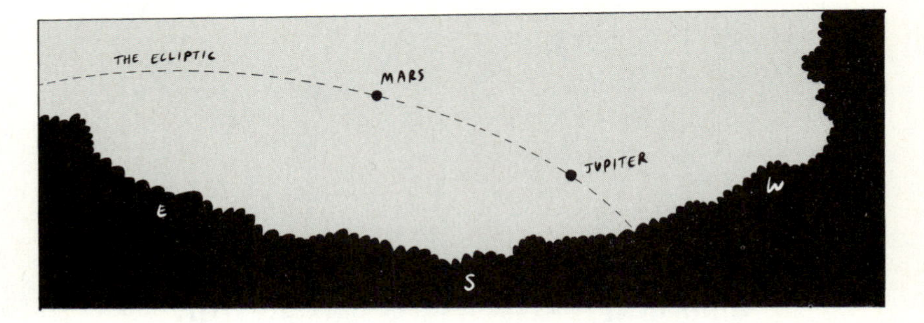

- A final thing to remember: make sure your drawing of the sky includes some constellations, so that you'll be able to notice the motion of the planets among the stars.

KEEPING IT SHORT

Simply identify which planets are visible in the night sky from your location whenever you decide to make your observations. In this case, you will not track the motions of the planets with respect to the stars over time but just try to locate them in the sky (as described previously).

THINGS TO PONDER

- Which planet or planets were most exciting for you to find? Why?
- Which of the planets you found seemed the brightest? Do you think you would have noticed that, or the difference between planets and stars, before doing this?
- As you view a planet outside at night, picture photons of light streaming from the Sun, reflecting off its surface or clouds, and then striking your retina. You are connected to that planet by photons of light every time you look at it. How does that make you feel?
- Have you ever looked at a planet through a telescope? What was your reaction to that moment?

MINDFULNESS EXERCISE (PRESENT MOMENT)

Planet Earth is covered in a life-giving atmosphere, made mostly of nitrogen and oxygen. Every moment of every day, this atmosphere flows around and envelops you. Close your eyes and try to imagine millions of molecules in the air right around you, bouncing off your skin, transferring heat into or away from your body. Be aware of the temperature of your skin. Is it warm or cool or neutral? Stay in the moment, feeling your connection to the planet.

THE MOON AND THE ECLIPTIC

The stars may not really move much relative to one another during a human lifetime, but the celestial objects closer to us—like the Moon—move a lot! Each of the planets bops across the sky at its own individual pace, following its orbit around the Sun and, in the Moon's case, its much faster orbit around the Earth.

The Sun moves against the background stars in its ecliptic path. In this exercise, you'll figure out what path the Moon follows, how quickly that path repeats, and how it matches up with the Sun's.

Noticing how the Moon moves can help you notice and appreciate the passage of time, because the Moon's cycles operate on human timescales, unlike much of the rest of the universe. And taking stock of how the Moon's motion stacks up against the stars can keep you mindful of its proximity, and aware of the fact that it's just out there, orbiting our Earth, all the time.

Although our Moon is, of course, special to us, in recent years scientists have discovered moons around planets outside our solar system. They have dubbed these orbs "exomoons." And so perhaps on some distant planet somewhere else in the universe, an alien observer is taking note of its own moon's motion through its own, different constellations.

WHAT YOU'LL NEED

A clear nighttime sky
Your Star Wheel or a night-sky app
A lunar calendar

This book, a notebook, or a sketchbook
A pencil or pen

WHAT YOU'LL DO

You'll watch how the Moon moves compared to the background stars, follow its journey through Earth's sky, and see how the trek compares to that of the very closest star, the Sun.

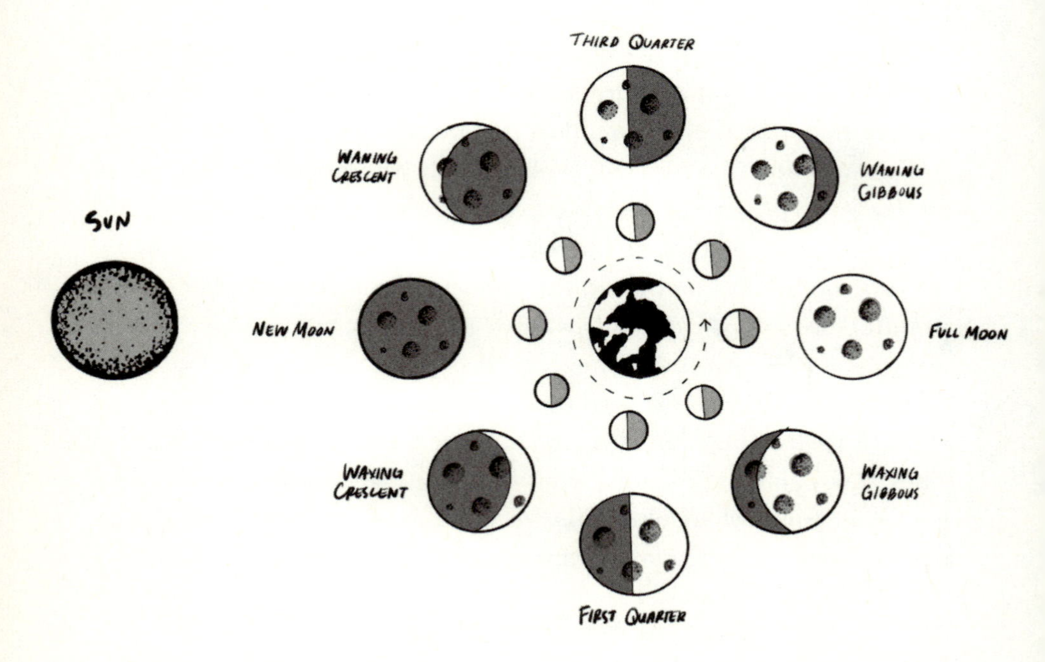

Phases of the Moon: this image shows the Sun, Earth, and Moon. As the Moon orbits the Earth (as seen from the North Pole), the Moon goes through its well-known phases.

THE DETAILS

Note that if the next steps seem like too much, that's okay! Skip ahead to "Keeping It Short."

- Using a lunar calendar, find out when the Moon will next be in its Waxing Crescent or its First Quarter phase (see the previous illustration of the Phases of the Moon).

- On that night, once it's fully dark out, go outside and view the Moon. See if you can tell which constellation the Moon is in. Use your Star Wheel or a night-sky app if they help! Write down in the space provided on these pages the constellation name and the date of your observations. If you feel like going all out, draw the constellation, its location, and the Moon's location. Create some landmarks or write down cardinal directions and approximate elevations from the horizon using your hand to orient yourself.

- Note where exactly the Moon seems to be relative to any nearby constellations (e.g., right in the center, slightly above, slightly below).

- Go outside about a week later, at about the same time, and see which constellations that big golf ball is near now. Write it down!

- Once you have a sketch (if you made one), draw an arc between the different lunar positions and any constellations you drew.

KEEPING IT SHORT

One night when the Moon is in its Waxing Crescent phase (see the Phases of the Moon illustration on page 28), figure out what constellation the Moon is in using pure observation, a night-sky app, or your Star Wheel.

THINGS TO PONDER

- Do you think the Moon and the Sun follow the exact same path through the sky?

- Do you think the Moon's path makes the constellations it lives in more important to humans, historically and presently?

- Can thinking about the constellations the Moon finds itself in help you mark the passage of its cycles—and so the regular passage of time—more mindfully?

- Take note of what the nearby constellations may have in common. (Hint: Are they constellations like Taurus, Scorpio, or Virgo?)

MINDFULNESS EXERCISE (BREATHING)

One straightforward way to let your mind relax and get into a meditative state is to count. It's the reason that counting sheep helps you to fall asleep. Go out at night and look at the Moon and count backward from thirty, taking a slow breath in and letting it out with every other number (thirty . . . in . . . twenty-nine . . . out). As you breathe, all you need to do is look at our planet's closest neighbor. How do you feel when you get to zero? You can replicate this even when the Moon isn't visible or you can't go outside—just close your eyes and visualize the Moon, counting backward, and breathing in and out.

WHICH WAY IS UP?

As far as astronomers can tell, on the largest scales, the universe looks the same in every direction and has no special locations. Wherever you might go in the cosmos, its properties would be the same. Wherever you go, there you are. So if the universe is the same in all directions, why are we spending time thinking about which way is up?

Because to consider the universe, it's helpful to orient yourself within it, like you figure out where your feet are before you get out of bed. This exercise will help you situate yourself in space using two coordinate systems astronomers have devised, one related to the Milky Way galaxy and one related to our specific solar system and planet. It can be a challenge to get your mind around what the solar system and the galaxy look like when you're stuck inside them. Think of it this way: If the Milky Way is a cookie, you're a chocolate chip inside that cookie. The Milky Way surrounds us in the same way that the cookie dough surrounds the chip, thicker in some directions and thinner in other directions! And the chocolate chip can't hop out of the cookie to peer down and simply see what it looks like, just as you can't zoom above the real solar system or galaxy to view them. The chocolate chip may only observe its surroundings.

WHAT YOU'LL NEED

A clear nighttime sky
This book, a notebook, or a sketchbook
A pencil or pen

WHAT YOU'LL DO

In this exercise, you'll learn about the coordinate systems that astronomers regularly use to locate objects in the night sky—like latitude and longitude, but in spaaaaace. A coordinate system is a grid map of the sky that we use to find our way around. Before that, you'll consider the coordinate system that comes most naturally: your local horizon and what's directly overhead. While local coordinates are handy when you're alone outside, they're less useful when telling others where to find a particular star in space, because what's overhead changes as the Earth rotates and depends on where you are. First, you'll find the line through the middle of the Milky Way, our home galaxy. Then you'll find the celestial equator, which cuts close to the middle of the solar system.

THE DETAILS

Note that this one is a little more involved. If you're skipping around in the book and haven't looked at the sky much, you may want to come back to this exercise!

Finding the Milky Way

This is a good exercise for the summer or fall. What you can see will depend on how dark your skies are. This is also a good exercise to try after you've become familiar with a few constellations.

- If you live far from city lights, you might be able to see the nearby parts of the Milky Way itself. At first, it just looks like a band of clouds. But real clouds will disperse relatively quickly, while the Milky Way will persist and rotate with the rest of the stars.

- If it's too bright at night to see the Milky Way from wherever you are, you can use the constellation Cygnus to locate the Milky Way. First, locate three bright stars that will be high in the sky soon after sunset in the summer and early fall. These stars are Vega, Deneb, and Altair, which together link up to form the Summer Triangle. For help, check out the Star Wheel (see appendix A, page 187). The star Deneb is located in the head of Cygnus the swan. The long neck of the swan runs right along the plane of the Milky Way. In the space provided, draw Cygnus.

- Above and below Cygnus, you should see the blended, fuzzy light of the Milky Way. You're looking through the middle of our galaxy! The Sun and about 400 billion other stars orbit its center. In the same way that our world is just one planet orbiting the Sun, the Sun is just one star orbiting the galaxy's core.

- Make an L shape with your hand, so that your thumb and index finger form a 90-degree angle. If you position your thumb in the Milky Way along the neck of Cygnus, your index finger will point out of this plane, into the vast emptiness between galaxies.

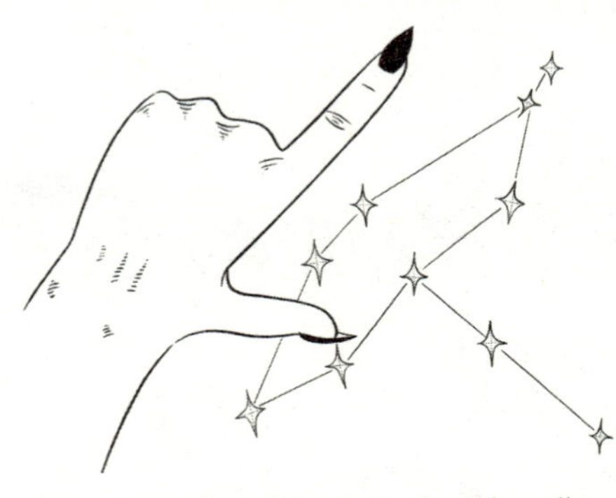

- Above and below your thumb, you're looking "out of the cookie," where there are fewer stars. Astronomers describe directions around the perimeter of the cookie as "galactic longitude" and directions above and below the cookie as "galactic latitude."

Finding the Celestial Equator

- Looking to the south, make a drawing of your southern horizon. Your drawing may contain buildings, trees, lampposts, or gargoyles. Whatever shapes are there, try to draw them carefully.

- Make a 90-degree angle with your two arms—a 90-degree angle just looks like the corner of a rectangle or a room. Keeping that 90-degree angle shape with both of your arms, point one arm at the North Star. If you've maintained that 90-degree angle, the other arm will now be pointing at the celestial equator, the invisible line you'd get if you pushed Earth's equator out into space.

- On your drawing of the southern sky, draw a dotted arc line that represents the celestial equator. It might look something like this:

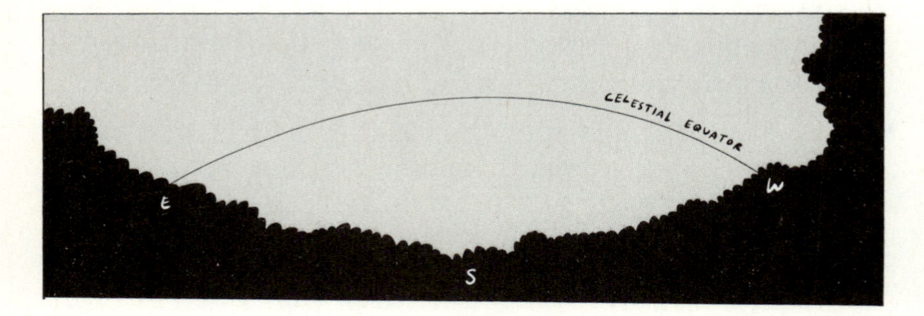

- You'll always find the Sun, the Moon, and all the planets in roughly the same part of the sky as the celestial equator. Which means you can use your drawing to hunt down planets, which will, at first glance, appear to be very bright "stars" in this part of the sky.

KEEPING IT SHORT

You can do this at any time, day or night. Go outside and stand in an area with a clear view of the sky. Orient yourself so that you're facing north (you can use a compass, a compass app, or any of the methods described in "Directions," page 16). Point straight overhead. That direction (otherwise known as "up") is called the "zenith." Astronomers sometimes refer to positions in the sky with what are called "altitude-azimuth," or alt-az, coordinates. The altitude of a position is the number of degrees something sits above the horizon. So something with an altitude of 45 degrees is halfway between the horizon and the zenith. The azimuth of a position is the number of degrees (clockwise) from due north, with east at 90 degrees, south at 180 degrees, west at 270 degrees, and 360/0 degrees back at north. Using this coordinate system, draw your local horizon in the space provided in one of the cardinal directions N (0), S (180), E (90), or W (270).

THINGS TO PONDER

- ⊙ Why might it be important for astronomers to have an agreed-upon coordinate system in the sky?
- ⊙ Do you think coordinate systems could ever be useful to you?
- ⊙ Is there any astronomical reason why the Northern Hemisphere should be on the "top" of a globe? Why do you think most maps show the Northern Hemisphere on the top half of the Earth?
- ⊙ What do you think would happen if, in your day-to-day activities, you paused more often to simply look up? What would you see? How would it affect your mood or your choices?

MINDFULNESS EXERCISE (CLOSE OBSERVATION/PERSPECTIVE)

Go outside on a clear night, when you can see at least a few stars. Keeping in mind that direction in space is arbitrary, imagine that the stars you see are not above you but below you, and that you're looking into the depths of a magnificently deep ocean. Allow yourself to witness how this thought makes you feel. Do you feel connected? Tiny? Alone? A part of the universe? Off-balance? However you feel, allow yourself to turn those feelings over in your mind and examine them without judgment.

INTERLUDE:
INDIGENOUS ASTRONOMY

Right below the grandmother spider is the Pleiades, the seven sisters. And that's called Pakone Kisik. The hole in the sky. And the hole in the sky is where we come from.

—WILFRED BUCK, ININEW/CREE FIRST NATIONS, CANADA[2]

When you search the night sky for constellations, you might recognize their names from Greek and Roman mythology: Orion the hunter, Andromeda chained to a rock, Pegasus the winged horse. Their stories come from Western culture, but every society connects their own forms and narratives to these same collections of stars—tales first told while their ancestors were gathered around flickering firelight. From Canada to Polynesia to Australia to Africa to the American Southwest, humans have long seen the night sky as a repository for their tales, as a set of guideposts for navigation, and as a way to gauge the passage and cycles of time.

The term "Indigenous astronomy" typically refers to these ways of knowing independent of, but related to, the story of astronomy in the modern West. Throughout this book are brief reflections on different Indigenous astronomy traditions and interviews with experts in these practices.

We are all descendants of the first *Homo sapiens*, some of whom left Africa approximately tens of thousands of years ago, reaching Europe, Asia, North and South America, and Australia across land bridges now submerged under big, sloshing oceans. Those early practices grew, evolved, and stuck around, remaining significant and useful to those who came after.

We are all connected to these early, keen observers of the night sky. Our hope is that you'll come to see modern astronomy and your own cosmic investigations as a continuation of these practices, and to appreciate these traditions as avenues to explore present-day mindfulness of the skies.

INTERLUDE:

PROFILE OF EARTH-MINDED SPACE SCIENTIST APARNA VENKATESAN

Astronomer Aparna Venkatesan grew up in India and Singapore, and went on to study the earliest elements, earliest stars, earliest gas, and earliest radiation that the universe produced.[3] But in addition to paying attention to the origins of everything in the cosmos, she pays attention, and wants science to pay attention, to the early and continuing knowledge of the night sky that Indigenous communities have produced. That's why she has for years worked to merge Indigenous knowledge with academic knowledge, and to include Indigenous knowledge holders at decision-making tables, in education and in research. That means, in part, not erasing anyone's identity or culture.

"The people I most love working with have really integrated their values with their work," she says. "Science prides itself on its rational, efficient approach to work, but that transactional way of doing things is limited in its outcome. For me, long-lasting change, meaningful work, comes from relational work. Indigenous knowledge is naturally a long-term, ecological thinking."

That philosophy is becoming more intertwined with the practice of science. "In Australia," for example, Venkatesan explains, "in a lot of disciplines you're required to have a slide in every presentation acknowledging the traditional custodians of the land—giving a nod to history but also noting that these people are still around."

In the US, missions and celestial objects now sometimes receive Indigenous or Native names, chosen by people from those communities, like the interstellar, cigar-shaped rock that, for a while, some people thought might be a spaceship. It's called 'Oumuamua, which means "scout" in Hawaiian. In general, says Venkatesan, "the question is really how much did we give complete autonomy in inviting them into this as opposed to having them come in at the very last stage to have the appearance of inclusion—which I think is an important first step but can't be the last step."

Scientists, Venkatesan has said,[4] should think of Indigenous knowledge as scientific, of oral traditions as the equivalent of peer-reviewed studies. Why? In part because Indigenous understanding is based on practical, long-term research that has proven itself for generations. "The way a community sets up its food chain, its food sovereignty, its fisheries, is tied to long-term sustainability," she says, "but it's also tied to sky practices. When do we fish? When do we plant? When your entire community's livelihood and sustenance are at stake, those practices are scientific. They let the community live for centuries or millennia."

It's important, Venkatesan says, to protect the skies so that humans can keep gathering knowledge, and using the knowledge they've been gathering for years and years. "I see dark skies as a human right," she says. "But I think it's especially important when traditional sky practices, sky traditions, or your cultural practices are dependent on dark skies."

Beyond that, the way Indigenous knowledge works could help us all in the increasingly complicated future. "As a planet and as a species, things are in crisis. I think we need partnerships with the global community, community-based ways of thinking, to move forward with solutions, because the planet has enormous pressing crises," she says. "It's almost reached a do-or-die point, to be sustainable and inclusive. To me, Indigenous knowledge is really about catalyzing the kind of creative change and creative solutions that we really need, right now, for the future. It's sustainable; it's interdisciplinary; it's all the things science says it wants."

She continues: "I see it also as humanity's heritage. I see it as humanity's experiential wealth. So why would we want to throw away or not partner with this knowledge?"

Photo courtesy of the University of San Francisco

APARNA VENKATESAN is a professor in the University of San Francisco's Department of Physics and Astronomy, where she does research in theoretical cosmology. She moved to the US to go to college at Cornell University. Her native language is Tamil, but she didn't stop there: she also speaks English, some French, and some Hindi, along with having been a singer for most of her life. In the US, she has worked to make astronomy more inclusive of Indigenous peoples and Indigenous knowledge, through both programs and policy.

II. THE SUN

IN WHICH YOU WILL OBSERVE SOME OF THE MOTIONS AND DETAILS OF THE CLOSEST STAR

You may think you know the Sun. After all, it's risen every day of your life—and set too. We all know it comes up in the east and goes down in the west. It's so reliable that we don't have to consider it, really, except maybe to be grateful for a sunny day after a wet week, or to notice that the days are waxing longer or waning shorter as the seasons progress. But what does any of that really mean? How does the Sun's position in the sky relate to your position on Earth, and Earth's position in its orbit around the Sun? And where exactly does the Sun set on the western horizon in late June? This section will help you make sense of your changing daily, monthly, and yearly experiences with Earth's own star, the Sun.

SUNDIALS

Most of us have phones in our pockets (or smart watches on our wrists) that tell us the exact time, to the minute and second. But that's not the only way to find out it's lunchtime. Other methods connect us with the sky, when we don't need to know that it's precisely 10:52:33 . . . 34 . . . 35. Sundials are one of the oldest tools (perhaps *the* oldest) to determine the time. The shadow of a stick can track the hours as they spin by, as they will on the sundial you'll make in this exercise, and remind you that our days only exist as such because there's a big fiery sphere millions of miles away giving us life and giving that life some structure.

WHAT YOU'LL NEED

An outside location in full Sun
This book
The sundial inserts (found in the back of the book)

WHAT YOU'LL DO

Making a sundial is a cinch. You could just push a very straight stick into the ground! As the Sun sweeps across the sky, the stick's shadow will drift from west to east. When the day is half over (you made it!), the shadow will point due north.

To up your sundial game, you'll need just a little more info: your latitude and the direction of geographic north. Cut out the sundial that most closely matches your latitude (there are two available in this book), and orient the sundial so the gnomon, or pointer, points due north. That's it. You're ready to go. You will have re-created one of the first clocks, which dates back at least three thousand years.

THE DETAILS

This book includes two sundials. One is to be used at 35 degrees north latitude, and the other is to be used at 40 degrees north latitude. (If you live somewhere farther north or south from there, you can customize your experience here: bit.ly/2JVcbPF.)

- Look up your latitude online.
- Using a compass (there's probably one in your phone—just open a map app), figure out in which direction north is.
- Cut out the appropriate sundial from the insert page in the back of this book.
- Fold the sundial along the dotted lines, so that the triangular part pops straight up above the sundial, then tape the sundial on the bottom so it stays folded. It should look like the picture on the next page.
- Lay the sundial on a flat surface in direct sunlight. A sunny piece of patio furniture or your office windowsill would be great. Rotate the sundial so the triangle faces north.

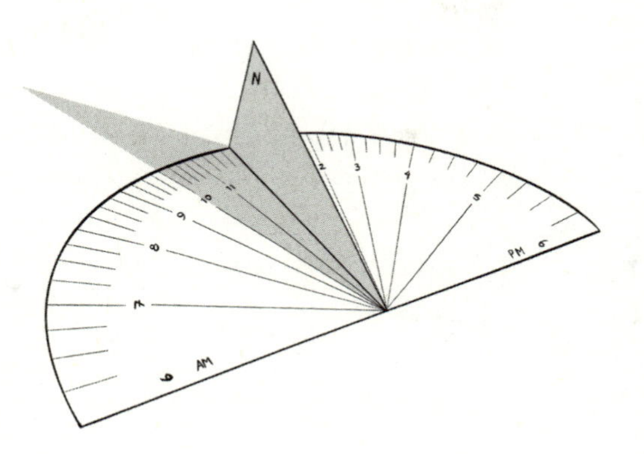

⊙ Throughout the day—and as the days pass from January to July to December—watch the shadow slipping along. Take note of what direction it moves, how fast it moves, and whether watching it change influences your experience of the day. Remind yourself that this clock works because there's a star not so far away and you're on a planet that's both curving around that star and spinning on its own, changing the way the light hits your particular location on Earth.

KEEPING IT SHORT

Without knowing your latitude, or the direction of north, you can simply push a straight stick into the ground in a sunny spot. Try to make the stick perpendicular to the ground—pointing straight up. Over the course of the day, from morning until night, notice the "sweep" of the stick's shadow across the ground. Which direction does the shadow move? What are the earliest and latest times you can observe? How does noticing the passage of time with this simple "shadow clock" influence your perception of time and its passing?

THINGS TO PONDER

- How does the sundial's measurement of time differ, in your experience of it, from looking at your phone or watch?
- How different would your life be if you only needed to know the approximate time, instead of the precise hour and minute?
- Many mindfulness experts talk about the importance of being present in the current moment. How does using a sundial help you with that?

MINDFULNESS EXERCISE (PRESENT MOMENT/CLOSE OBSERVATION)

We often rush through the days, meeting deadlines and refreshing the calendar, which just keeps pinging with alerts, constantly. *Am I late? Am I early?* Many of us have gotten to the point when if we have a spare moment, we scroll through social media. The next time you arrive for a dentist appointment early, or meet with an ungodly long checkout line, make a conscious choice not to take out your phone. Use those thirty seconds or three minutes to look around you, noticing the other people in line or in the waiting room. Visually take in your surroundings and notice the space. Be aware of this present moment.

WHY ARE THERE SEASONS?

The seasons come and go, and as they pass, you might notice that the Sun seems to beat down on you more during summer months, while the air has more of a chilly snap, even midday, in the winter. It turns out, there's a good reason for these changes, and an explanation for why winter is summer's fraternal twin if you live near the equator. These changes have a lot to do with the changing position of the Sun at midday.

Earth is tilted on its axis, leaning at 23.5 degrees currently (it has varied between 22.1 and 24.5 degrees over the past 40,000 years). Some long-past collision, perhaps the one that created the Moon, might have knocked our planet a little to one side. As a result, where the Sun rises and sets—and how it arcs between the two places—changes throughout the year (see "Position of the Setting/Rising Sun," page 55). The following drawing shows how the Sun's path in the sky shifts as the

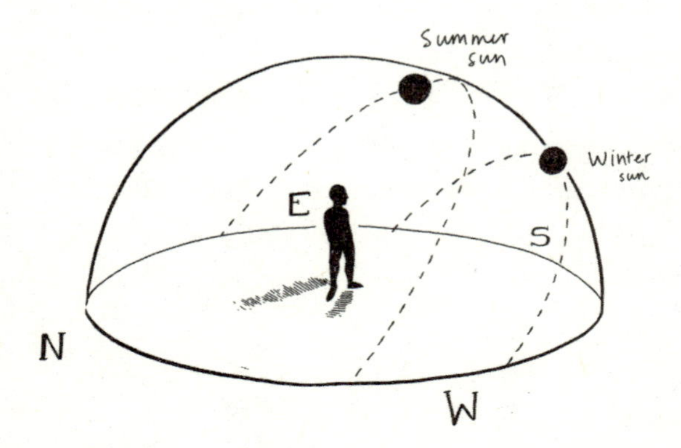

The Sun travels a higher path across the sky in summer, and a lower path in winter.

calendar flips from winter to summer. Note that in the winter, the Sun rises farther southeast, and it's lower in the sky and never right over your head. In the summer, the Sun is above the horizon longer and higher up, giving us those lovely late-light evenings in June.

WHAT YOU'LL NEED

A clear daytime sky
This book, a notebook, or a sketchbook
A pencil or pen

WHAT YOU'LL DO

During the middle of the day, in different seasons, you'll watch to see how high above the horizon the Sun crests, using humans' oldest and probably best tool—your hand—to measure.

THE DETAILS

This procedure assumes that you live in the Northern Hemisphere. If you find yourself down under, then you'll measure the angle between the Sun and the northern horizon.

- Go outside on a clear day and try to choose a place where you have a reasonable view of the southern horizon and the Sun. Try to pick a time close to the middle of the day, maybe when you skip out for lunch.

- Make a note of the date and time.

- Using your hand, measure the approximate angle between the southern horizon and the Sun (see the diagram below). Don't stare into the Sun as you do this—it's a bad idea regardless of what day or time it is! Make a note of the angle or even just how many fingers fit between the Sun and the ground.

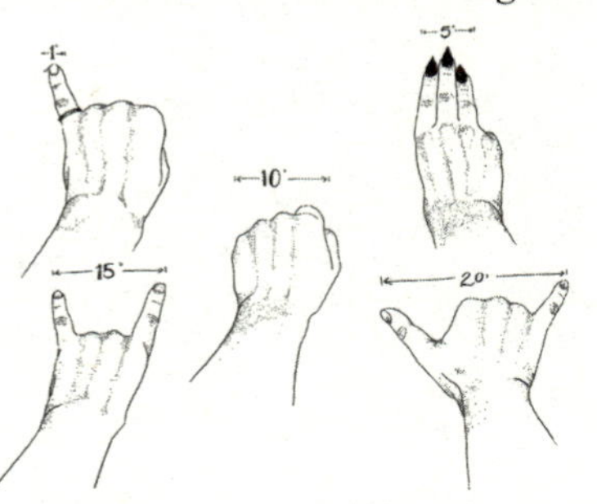

- Try doing this once a month, or whenever you happen to think of it. Write down your measurements.

- How much of a difference exists between your most-different measurements? It's easier to understand why days are shorter in the winter, and why the Sun feels like a colder star in December than in June, when you notice that it's truly beating down on you in one season and just glancing its light over you in another.

KEEPING IT SHORT

Make just two observations of how far the midday Sun is above the southern horizon, one in December and one in June, when the Sun sits at its two most extreme positions in the midday sky.

THINGS TO PONDER

- Do you live in a place where the summer and winter are very different in terms of temperature?
- Do you enjoy the change in seasons? Why or why not?
- What's your favorite season, and does it have anything to do with the Sun?
- How does it make you feel to use your own body (your hand) to measure astronomical distances?
- Have you ever noticed that the Sun is directly overhead? And that you have no shadow?

MINDFULNESS EXERCISE (PRESENT MOMENT/CLOSE OBSERVATION)

Go outside near noon on a sunny day. Find a place where you can be away from others, and sit comfortably on a wall, bench, or the ground. Close your eyes. Notice what the sunlight feels like on your skin. Focus on the sensation of the sunlight warming your body as you sit there, absorbing its energy from 93 million miles away.

POSITION OF THE SETTING/RISING SUN

One star rises and sets every single day and has done so since before Earth even got its first rudimentary microbes. You may have heard of this star: the Sun. Before humans had alternate sources of light, they used to rise and set themselves along with it.

But once we invented candles, oil lamps, and light bulbs, daily routines became pretty disconnected from the routine of our star. Because of that, we don't tend to notice it much from day to day. Sure, we know that in winter its position creates shorter days (sometimes so short that it's only light when we're at work), and in the summer the Sun stays out late for backyard cookouts. Beyond that, though, the details can feel fuzzy.

Usually, in the morning and evening, we're on autopilot: get ready for work, go to work, return home, walk the dog, hang with the kids, make dinner. The daily beginning- and end-of-day habits happen without much thought, and while we're engaged in them, we don't tend to focus on what's going on around us. Like the fact that a star—a giant fusion reactor—is spraying out radiation both life-giving and deadly, and arcing through the sky above our heads.

Taking a few minutes to disengage from autopilot and to engage with that star can be like pressing a reset button, refocusing on your present environment and its evolution.

But making time and expending effort to notice an orange circle in the sky may not come naturally, so try this to help yourself: record the position of the rising or setting Sun relative to local landmarks. All you need is a clear view to the east if you're getting up early or to the west if you're a night owl. If you live in a city, this exercise might be easier from a rooftop or a park you pass on your commute. Choose anywhere that you can have a relatively unobstructed view of the horizon.

Local weather will vary, you might take a trip, you might forget, or you might have to stay late at work. Your observations may not be perfect, and that's okay. Try to observe the Sun without judgment of yourself or your ability to do this as often as you "should." Just make a map of the surroundings at your chosen position (by the way, this drawing can be a good way to be more mindful of your local environment), and mark the position of the setting Sun on this map whenever you can. Within a few weeks, even if you can't make it outside often, patterns should start to emerge.

WHAT YOU'LL NEED

A clear daytime sky
This book, a notebook, or a sketchbook
A pencil or pen
A couple of minutes every few days

WHAT YOU'LL DO

Find a good viewing position to look either east, toward the rising Sun, or west, toward the setting Sun. If you've got a flexible schedule, why not try both directions? Toward the end of this exercise, there's space for you to draw your local horizon, or you can do so in your own sketchbook. Notice and draw everything you'd normally screen out: buildings, lampposts, trees, street signs—anything that will help you carefully record the Sun's location. It will also be helpful to know the rough "angular sizes" of all that stuff: how big they look in the sky. Knowing angular size requires no special tools or expertise—just your hand. For reference, the Full Moon has an angular size of about 0.5 degrees (or about half of your pinkie finger held at arm's length).

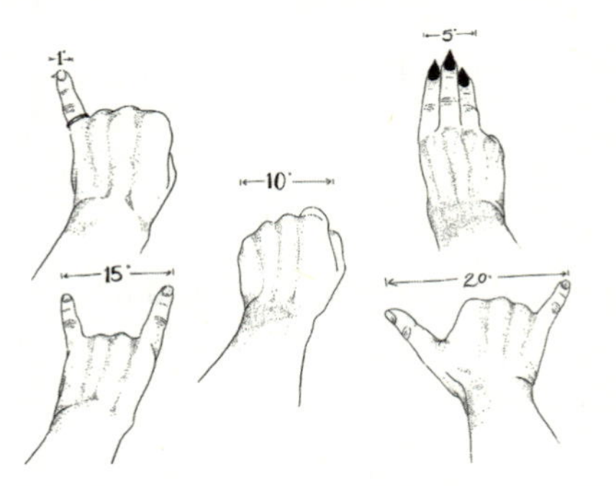

When held at arm's length, your hand can be conveniently used to roughly measure angular size. The numbers above each sketch (in degrees) indicate how much of the sky is covered. For comparison, the whole sky, from horizon to horizon, is 180 degrees.

After a few weeks of sketching the Sun, you'll notice what a journey it's been on, and then you can make an educated guess about where—relative to all those newly noticed trees and buildings—it will be in the future. Congrats: you're a cosmic fortune-teller.

THE DETAILS

Note that it would be best not to start this exercise in mid-December or mid-June, when the Sun reverses direction on the horizon. That will complicate your life unnecessarily, and nobody needs more complication. Also, these are instructions for a western horizon at sunset—just copy and paste them in your brain for the eastern horizon at sunrise!

- Find a place where you can see the horizon to the west, maybe perched atop your roof, cross-legged on the hood of your car, or wrapped in a blanket in a nearby park. If you have lots of high buildings or trees around you and your view is blocked, just plan to do your Sun watch about twenty minutes before the official sunset time, when it will still be a wee bit above the horizon.

- ⊙ In the space provided on the next page or in a sketchbook, draw a picture of that horizon complete with landmarks. Take time to notice things your eyes might otherwise overlook—trees, your neighbor's chimney—but don't bother drawing things that are likely to change position, like Karen's recycling bin across the street.

- ⊙ Figure out what time the Sun will set on the day you will do your Sun watch. Devices make this easy: If you type "time of sunset" into a search engine, it will magically tell you. So will a smart watch and most weather apps.

- ⊙ Around sunset, go to your special spot and look to the western horizon.

- ⊙ Don't stare at the Sun.

- ⊙ On your drawing, mark where the Sun is setting relative to all those chimneys and trees. Write the time and the date next to your observation.

- ⊙ Repeat your observations whenever you can.

- ⊙ As time passes, take a moment to notice whether the Sun has moved relative to the lampposts, apartment buildings, and oaks in your drawing. You know why that movement happened? Because you and this whole planet are moving through space, and Earth's axis is a bit tilted. As a result, throughout the year, the Sun's position at sunset slides back and forth on the horizon.

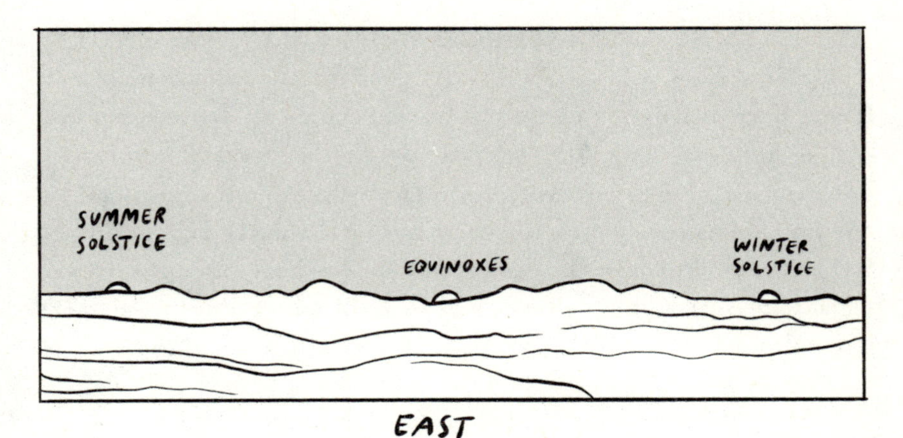

The Sun moves back and forth on the eastern horizon at sunrise through the course of the year.

⊙ After you've spied on the Sun for a few weeks, you could use that knowledge to guess where it will set an equal number of weeks in the future.

KEEPING IT SHORT

If you have a camera on your phone, you can do an abbreviated version of this exercise: Wherever you are, at whatever time of year, use a compass (or a compass app) to face west around sunset. Take a picture of that average star at sunset, as low in the sky as you can and from the same place every time. After two or three months, collect your images in a single album, and look through them in order. It's like a flip-book!

THINGS TO PONDER

- Have you ever gone camping and noticed how tired you get once the Sun sets, without the barrage of artificial light? How would your life change if it were like that all the time?

- You can observe the passage of time, and the switching of seasons, just by noticing what the Sun is doing. If you watch the Sun over time, eventually it will stop moving in one direction and start heading in the opposite direction. That happens in late December and late June, on the solstices. The Latin *solstitium* translates as "Sun stops." Why might knowing about the solstice have been useful in the past? How and why do you think modern society, and you, have become disconnected from that?

- There are two days when the Sun will set not just somewhere in the west but due west. Those days are the spring and fall equinoxes. Noting the equinoxes was useful to societies in the past, but it's fallen out of favor. Why don't we need to note them anymore, and what might change if we did?

- Have you ever noticed that light comes into your house through a particular window in an annoying (or at least noticeable) way, but only some of the time? This happens because the Sun doesn't travel the same path in the sky every day but a slightly different one!

MINDFULNESS EXERCISE (CLOSE OBSERVATION)

You'll be drawing the position of the Sun over a period of time, becoming deeply familiar with the shapes, colors, and locations of objects on the horizon that you're watching. As time passes, try to notice changes in the objects hanging out on the horizon. If there is a tree, have its leaves changed in color or abundance? If a mountain is on that horizon, have its colors changed subtly? Has that house across the street gotten a new paint job? Before carefully drawing the Sun's location, take a few moments to explore your surroundings, noticing subtle changes as the seasons pass.

THE DISAPPEARING STAR

The Earth, the Moon, and the Sun are in a strange and lucky arrangement. From the vantage point of the planet's surface, our nearest neighbor (the Moon) and our closest star (the Sun) both appear the same size in the sky. That's just a coincidence. And once in a while these three spheres line up with one another: the Sun shines on the Moon while Earth is directly behind it, blocking the Sun from our view, casting the Moon's shadow on part of the Earth and causing a solar eclipse. It's like if a friend held a basketball up to precisely block a spotlight.

When this alignment happens, the shadow moves across Earth, casting some areas into eerie darkness for a few minutes as it contin-

TOTAL SOLAR ECLIPSE

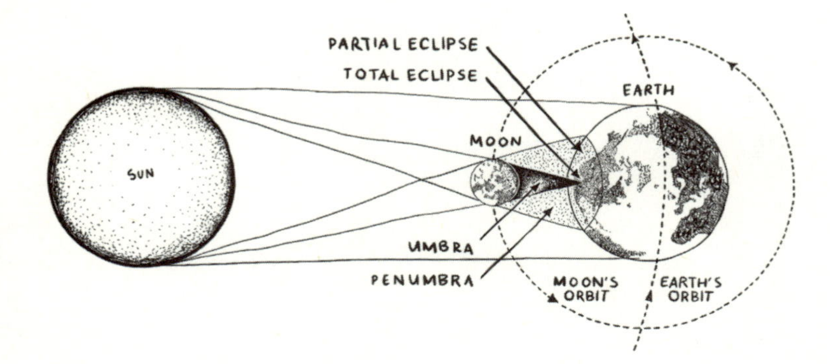

EARTH AND MOON DIAMETERS TO SCALE, SUN DIAMETER AND DISTANCES BETWEEN SUN, EARTH, AND MOON NOT TO SCALE.

ues its path across the planet. If you're lucky or resourceful enough to be in the path of a total solar eclipse, prepare to have your mind blown. The day will turn dark; the temperature will drop; crickets and frogs will start to chirp; you'll be able to briefly look directly at the Sun. The Sun's outer atmosphere, its corona, will appear for perhaps the first time in your life.

For some, a total solar eclipse is a profound experience: Earth becomes a whole new, strange place. Seemingly the most constant thing—that the Sun looks the same as it moves across the sky every day—isn't quite true anymore. And in that moment, when the "solar" part of the solar system is blocked by *another* body in the solar system, it's much easier than on any other day to feel like you live on an actual planet in actual space. Once the event is over, that knowledge and sensation usually remain.

Take note! Seeing a partial eclipse is *nothing like* seeing a total eclipse. A total eclipse is exponentially more amazing, so if you saw a partial eclipse and thought, *Eclipses are no big deal!* think again. And next time, try to get to the "path of totality."

WHAT YOU'LL NEED

Eclipse glasses, a colander, or the dappled shade under a tree
This book, a notebook, or a sketchbook
A pencil or pen

WHAT YOU'LL DO

You'll watch the Moon cross paths with the Sun, both blocking it and revealing more about its true nature. The idea is to focus on your experience of this event and to reflect on it afterward.

THE DETAILS

- Find the next solar eclipse that will occur near you (it may be a while—for example, the next solar eclipse visible in North America will not happen until April 8, 2024). Check out a website like timeanddate.com, which has a page listing future eclipses. The dates and approximate locations of total eclipses expected in the next decade are also shown at the end of this exercise.

- At the appointed day and time, either don your eclipse glasses and start watching or use one of these two contraptions:

    - A colander: You can simply hold it up, and its holes will project a bunch of miniature Suns onto the ground for you, like stellar polka dots.

    - A tree (any tree with leaves!): Before an eclipse, the light shining through the leaves will look like little blobs, but those are actually projections of our star! As the Moon begins to cover the Sun, the blobs will slowly turn into crescents. It helps if the ground under the tree is relatively smooth (think concrete, not grass).

- Even if you can't see a solar eclipse from where you are, you might catch a partial one—the same methods already listed will work for that too, although the final effect will be less dramatic!

- In the lead-up to "totality," as the Moon invades the Sun's space, watch as the Sun gets a disk shape cut out of it, like a cookie with a bite taken from it, and that bite grows bigger.

- Notice everything you can about the changes in the environment—temperature, animal behavior, brightness, wind. Focus on experiencing it so you can remember, and perhaps record, it later. Watch for an effect called the Diamond Ring, right before totality, when the only sunlight visible shines against the rugged craters at the Moon's edges.

- When the Moon totally covers the Sun, take off your glasses. You can look right at the closest star! Notice everything you can about how it looks—because chances are you won't be able to stare at it again, at least not for many years.

- Notice too how it makes you feel. People have a wide range of reactions—from yelling spontaneously to crying to feeling scared, even though they're not in danger, to feeling giddy.

- When totality ends, you can continue to watch with your eclipse glasses, colander, or tree as the Moon's shadow slides away and the Sun goes back to normal—even though you'll probably never see it the same, normal way again.

- When the eclipse is over, and you've had time to process your experience, consider writing down an account of it. It's often hard to remember what out-of-the-ordinary experiences truly felt like when you're back to being ordinary, so you might like to have an account of it to look back on later. Sketch the Sun at totality in the space provided, if you're feeling visual. If weather did not cooperate in your location (clouds can be a bummer), you can find and watch a video recording of the event made by an observatory somewhere else along the path of totality.

KEEPING IT SHORT (AND TRAVELING LESS)

Search for images or video online of solar eclipses past and present. Notice differences between them, based on where they were taken and what the paths across.

THINGS TO PONDER

- Have you ever experienced a total eclipse? Do you remember how you felt?

- What would you have thought of a solar eclipse if you didn't know what was happening? Do you think you'd find it awe inducing, terrifying, or both? Why? How would you have explained it to yourself?

- What do you think ancient people thought about eclipses?

- Some people become so enamored of these events that they become "eclipse chasers," traveling the world just to catch a glimpse. Would you ever do that? Why or why not?

MINDFULNESS EXERCISE (CLOSE OBSERVATION/PRESENT MOMENT)

During one day, take a moment once an hour—or whenever you happen to think of it if you don't want to be that rigid—to notice the quality of the sunlight and the sky. What are the dominant colors? How bright is the sky compared to an hour before? How has the qual-

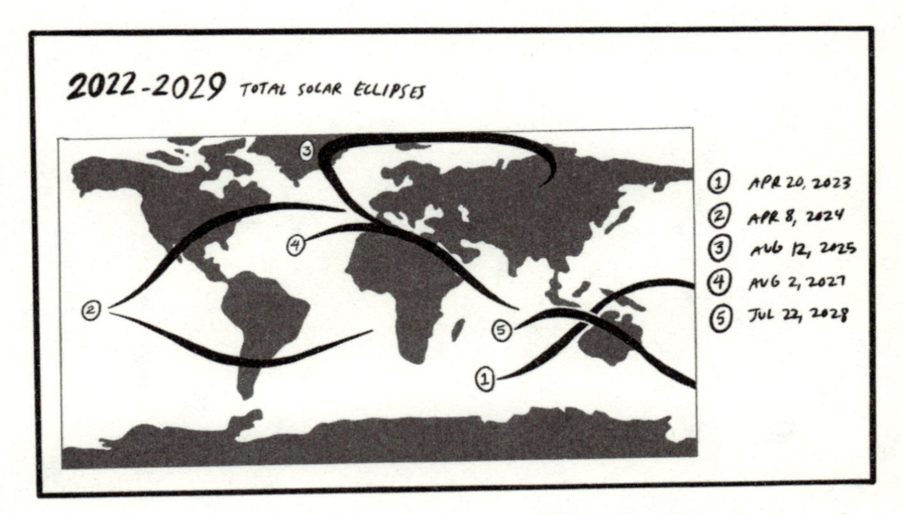

The slug-like shapes indicate the paths across the Earth's surface where a total eclipse will be observable. Partial eclipses will be observed outside those areas.

ity of the light changed? Are there sharp-edged shadows? If you're near a window or outside, can you feel the Sun on your skin? Take stock of how the light—whether it's gloomy or dappled or brilliant—matches up, or doesn't, with your mood. Stepping back from whatever you're doing or thinking to notice what's around is a mindfulness practice you can use to experience the moment and then refocus on whatever other task is at hand.

INTERLUDE:

THE ANCIENT ASTRONOMY OF AFRICA

African history is more complex than people would like it to be.

—THEBE MEDUPE, ASTRONOMER[1]

The Aboriginal people of Australia and the residents of North Africa made some of the earliest known celestial observations.[2] African contributions to astronomical knowledge and understanding—both historical and recent—were highlighted in the film *Cosmic Africa*. This documentary tells the story of Thebe Medupe, a South African astrophysicist who, from an early age, was curious about Africa's role in celestial understanding. The film highlights several regions and stories that shed light on Africa's long history of cosmic awareness.

Nabta Playa, for instance, is an arid region in southern Egypt that appears to be the home of some of the earliest astronomical constructions. Dating from about 4800 BCE, these beat the construction of the more famous Stonehenge by about two thousand years. Here, four massive, shaped stones sit, arranged so that two are oriented north–south. North is the direction about which the night sky appears to rotate as the Earth turns. The other pair of huge stones align with where the Sun rises on the longest day of the year, the summer solstice. To this day in the Nabta Playa, the summer solstice is followed by life-giving monsoon rains.

Cosmic Africa also features the Dogon people of Mali. The Dogon settled this region about five hundred years ago, and tied the motions of the stars and the Sun to agricultural work. The Moon's travels tended to relate to social or religious activities. In a world where many of us fend off nightfall with the lights and sounds of our interior spaces, the Dogon have a very different experience—the one most Westerners only occasionally savor when camping or watching a meteor shower.

The Dogon dwellings in the Bandiagara Escarpment (Mali) are reminiscent of the cliff dwellings in Mesa Verde, Colorado, ancestral homes of the Pueblo Indians.[3]

In Namibia, the Ju/'hoan people remained isolated from the rest of the globe well into the twentieth century. As a result, many of their astronomical traditions survived, including paying close attention to the Milky Way, which the Ju/'hoan people call the "Spine of the Night." The changing position of the Milky Way and its repeating patterns are a way to mark the passage of time and to determine when to begin particular activities. A study of the Ju/'hoan and their first use of fire makes an interesting conjecture: fire allowed activity to continue into the evening, so when people gathered around firelight, religious and cultural activities began to flourish. The night sky provided a constant framework for reflection, and people used it as a repository for stories and, in some cases, laws.[4]

With the spread of Islam in North Africa after around 800 CE, close observations of lunar phases and interest in cardinal directions—essential for determining the direction of Mecca at prayer time—grew in importance. Because the New Moon determines the start of each month in the Islamic calendar, lunar-phase tracking fit well with practices that were already an important part of life in much of North Africa. As anthropologist Christian Vannier puts it, "Non-seasonal, lunar-based Islamic ritual calendars were easily mapped onto local agricultural and ritual calendars."[5]

Africa's contributions to astronomical knowledge continue to the present day with the construction in South Africa and Australia of the Square Kilometre Array (SKA), which will be one of the most sensitive radio telescopes in the world.

INTERLUDE:

PROFILE OF FELLOW STARGAZER THEBE MEDUPE

"I was thirteen years old and Halley's Comet was visible in the night sky," says Thebe Medupe, a professor of astronomy at the University of Cape Town in South Africa, of how he first became interested in celestial matters.[6] "We had a very nice library in my town, and I went to my library, and I found books on astronomy." One book detailed how to build a telescope with mundane materials Medupe could get from his school.

"I went and I built my telescope, and I pointed it at the Moon," he says. "I couldn't believe what I was seeing. I then pointed it to Jupiter and saw all the moons of Jupiter. From then, my interest in astronomy and science just grew and developed until I was able to go to university."

Medupe became one of South Africa's first Black astronomers, and today he researches stellar shaking. "I study seismology of stars," he says. "You look for seismic waves in stars. You take them and measure their properties, from which you can infer the physics of stars. The seismic waves manifest themselves in the light output of a star. So a star with seismic waves will have light that is not stable—that goes up and down. And by measuring that light and that periodic behavior, you can determine a lot of information about the physics of stars. Whether the star is stable or not. Whether the star is vibrating or not. You can determine the interior composition of a star, even."

It is that inside view—accessing the hidden parts of distant suns—that draws Medupe to his chosen field. "You can measure properties of stars that you wouldn't be able to if these stars did not have these seismic waves," he says. "If you look at a star with a telescope, you have access to only the very thin outer layers. If you measure seismic waves, these waves travel throughout the *whole* of a star. They offer a window into stars."

But Medupe's scientific interest hasn't just extended upward. It also has extended outward, to other Black people who might want to take an orbit similar to his but just don't know it yet—or know how. "South Africa became a new country in 1994," he says. "At the time there were no Black astronomers in the country. Being a Black astronomer, I wanted to make sure that things changed and that we had more and more Black South Africans, and Black people in general, doing this as a career." To figure out how to bring others in, he met up with a group of South African astronomers. "Together, we formed a consortium called the National Astrophysics and Space Science Programme, where we brought astronomers from all across South Africa to run the program."

Although Medupe has passed that particular torch on, and others now operate the program, it continues to do its job. "The numbers have gone up quite a lot," he says. "Because of the program, which was funded by the South African government, we have many astrophysicists not just in South Africa but also in Uganda, Ethiopia, across Africa."

It was never just about training more African astronomers, though: it was also about revealing to people in Africa—and across the world—that the continent has a long and storied history of studying the universe.

That's part of why Medupe became involved in the *Cosmic Africa* film. "Around the time I was doing my PhD at the University of Cape Town, two filmmakers were thinking of making a documentary on African astronomy—cultural astronomy, mostly. They wanted to build a story around a character. At the time, I think I was one of the two Black people in the country who had a PhD in astrophysics."

The documentary was a discovery process, in which Medupe traveled around the continent, investigating different astronomical traditions and gaining knowledge—about how bushmen interpreted a solar eclipse, how the Dogon people integrated the stars into their daily lives—that had been erased in the modern political moment. "During apartheid, African science didn't make any mention of African scientists or African heritage. Really, our knowledge of the history of science in Africa was very limited."

One of the best parts of the experience was seeing how Africans synthesized knowledge from different disciplines. "The difference between Western knowledge and African knowledge is African knowledge is integrated, whereas Western knowledge is all categorized into little boxes. And sometimes even students struggle to see the link. Math is taught in the department of mathematics, physics in the department of physics. Really, all these things are entwined."

That project led Medupe to another film, *The Ancient Astronomers of Timbuktu*, about Timbuktu's vast archive of manuscripts. "There were hundreds, thousands of books all over the city in people's private collections, and lots of material—books on different topics, some on astronomy, some on mathematics, some on Islamic studies. We translated about thirty of these books on astronomy and mathematics, and we published some," in addition to detailing them in the documentary. Medupe recalls, in particular, a method scholars had for calculating the leap year using the Islamic calendar. "I wrote a little program based on the algorithm," says Medupe. "It was just beautiful to see that the way to calculate this was in this book from three hundred, four hundred years ago."

Medupe's work has shown the world the complicated history of the African continent—something that he knew but that had been obscured by politics and racism. And seeing through to that truth, looking at how its waves propagate, tells a whole story, not just one about the outer layers.

THEBE MEDUPE was raised in Mmabatho, a city in the North-West province of South Africa, and once the region's capital. Medupe is currently an astrophysicist at the University of Cape Town, where he continues to study the oscillations within stars. He leads a project to archive, digitize, and analyze scientific records from Timbuktu, and was the lead in the documentary *Cosmic Africa*. His work in astronomy outreach has resulted in an increasing number of Black African astrophysicists, schooled in their continent's and countries' scientific history as well as the continent's modern work in astronomy.

III. THE MOON

IN WHICH YOU OBSERVE SOME OF THE MOTIONS AND DETAILS OF OUR CLOSEST CELESTIAL NEIGHBOR

SCULPTING THE MOON*

SKETCHING WHAT YOU SEE IN THE MOON*

OBSERVING A LUNAR ECLIPSE*

WHY DOES THE MOON CHANGE?*

THE UPS AND DOWNS OF THE OCEAN***

For thousands of years humans have linked Earth's closest neighbor to their own behavior. People have connected the variation in the Moon's twenty-eight-day cycle to mood, violence, fertility, and even supernatural powers. While those connections don't always hold up to scrutiny, our planet's satellite does matter to the presence, persistence, and processes of life on Earth. Scientists are still debating exactly how it formed, but regardless of how it got here, once it existed, the Moon began to tug on Earth. The Moon causes tides, stabilizes Earth's axis, keeps seasons consistent, and influences animal behavior. Doodlebugs, for instance, dig bigger holes in the sand during a Full Moon. And yet it may be easy for you to go about life largely without noticing the Moon, until one day you're driving home from work and it's low and orange on the horizon, or it shows up above the office at noon, and you think, *What's the Moon doing?*

With the following exercises, you'll watch as the Moon progresses through its phases and discover how its alignment with Earth and the Sun causes these changes, how its gravity shows up in your everyday life, and what its size and distance actually mean in practical terms.

Although the Moon is the only other solar-system body humans have walked on, it remains mysterious and the place astronauts are most likely to venture to in the near future. By noticing and understanding how the only object within a few days' "ride" relates to your life, you'll realize anew that life here on Earth is connected to what goes on in space.

SCULPTING THE MOON

Maybe you remember when the Apollo 11 crew traveled to the Moon, then set foot on it for the first time. Maybe you've pulled up a clip on YouTube and seen thick-suited astronauts move slowly down the ladder from their lander to the lunar dust. Or maybe you watched the 2018 movie, made from that archival footage, about the trip.

Regardless, that trip, and the few that followed before the Apollo program ended in the early 1970s, provided the only concrete sense that humans have of how big the Moon is and how distant it is from us. Other than those missions—and photos that involve no humans—our closest neighbor remains a bit abstract. It feels like a flat gray disc in the sky, not a cracked and cratered world, spinning around the Earth, traveling through space.

It took Earth's earliest space trekkers three days to get to the Moon. Three days of cruising through empty, silent space, as the Moon grew larger and larger and Earth grew smaller and smaller. Perhaps you've seen the photo called *Earthrise*, taken from an orbit around the Moon during Apollo 8, which shows our planet hanging above the lunar surface, like the Moon hangs above Earth's—a literal change in perspective.

But how big a world is the Moon? And how distant? In this exercise, you'll make an educated guess, and then make your own Moon-Earth system, sculpting it from Play-Doh (or whatever squishy substance you have), estimating how large they are compared to each other, and how far apart.

WHAT YOU'LL NEED

Play-Doh or clay or even pie or pizza dough
(see box for a modeling dough recipe, page 82)
A knife
Two plates

WHAT YOU'LL DO

You'll sculpt both an Earth and a Moon, trying to make them the same relative sizes they are in real life. Then you'll estimate how far apart they would be if they were the size of your sculpted globes.

THE DETAILS

- Take whatever squishy substance you're using from its container.
- Using *all* the material, create an Earth and a Moon that you *think* represent their sizes relative to each other. For example, if you think the Moon's volume is a quarter of Earth's, cut the dough into five equal pieces, squish four of the pieces together to make Earth, and roll the remaining one into a ball for the Moon. Write down your guesses.
- Get creative if you feel like it—you can scratch continents and craters into the two spheres with your knife.
- Now put each of your globes on a different plate.

- ⊙ Set the Earth plate on the ground.
- ⊙ Move your Moon plate however far away you think the Moon would be from Earth if Earth were actually this size.

KEEPING IT SHORT

Draw an Earth. (Or, if you want to keep it super short, draw a circle!) Now, imagining Earth is that size, draw a circle corresponding to the Moon. Write down a guess for how far apart they'd be.

THINGS TO PONDER

- ⊙ Spoiler alert!! Here are the right answers: The Moon's diameter is about a quarter of Earth's. But that means the Moon is just 1/49 of Earth's volume, so if you cut the Play Doh into fifty pieces, forty-nine become Earth. The two are about thirty Earths apart. Another way to wrap your head around it: if the Earth were your head, the Moon would be around thirty feet away from it.
- ⊙ Without looking at the Moon in the sky, how big would you guess it looks off the top of your head? As big as your fist held out in front of you? As big as a golf ball held out in front of you? Next time you see the Moon in the sky, check it out. To most people, it is surprisingly small!
- ⊙ Is the actual distance between Earth and the Moon hard for you to comprehend? Why or why not?

- Do you think it would matter in your life if the Moon were bigger or smaller in the sky than it is?

- The next time you can see the Full Moon, go check it out. What would it have been like to put the first footprint in its dust? How do you imagine that felt?

- Do you think the astronauts were mindful of their presence on the Moon?

Play-Doh Recipe

Ingredients

1 cup flour
¼ cup salt
¾ cup water
3 tablespoons lemon juice
1 tablespoon vegetable oil

Put the wet ingredients in a saucepan and heat them until they're warm but not quite boiling. In the meantime, mix the dry ingredients together in a bowl. When the liquid ingredients are warm, stir them into the dry mixture. The dough should begin to clump up and cohere. Voilà!

Frederick County Public Libraries
Urbana
06/05/2023 04:10:27 PM

Customer Name: GADD RAG
Last 4 Digits of Customer ID: 4524

ITEMS BORROWED:

Title: Astronomical mindfulness : your cosmic guide to reconnecting with the sun, moon, stars, and planets
Call #: 523.2 DEPR
Item #: 21982320225380
Due Date: 06/26/2023

--Please retain this slip as your receipt--

MINDFULNESS EXERCISE (VISUALIZATION)

Visualization, rather than narrowing the mind in on the experience of external reality, focuses it on a scene you create inside your own mind. Close your eyes and imagine you're the only person inside a small spaceship, just leaving Earth. You're cocooned against the harsh outside environment, traveling toward the Moon at a constant velocity and in zero gravity. The Moon grows larger and larger through your tiny window. When worries or concerns occur to you, examine them and then let them go. Picture your concerns drifting behind you as your spaceship continues its journey toward the Moon (your worries don't have a rocket booster and can't keep up) and you leave Earth behind. Keep traveling until the Moon fills your view.

SKETCHING WHAT YOU SEE IN THE MOON

The Moon, like the dust on your windowsill or your coat hanging by the door, is pretty much always around. And so, like the dust and the coat, it's natural not to look much at it. Or to look at it but not really *see* it. Human nature leads you to dismiss details you don't need, or things that remain essentially the same all the time, because there are so many other sensory stimuli to take in and process. It's the same reason you eventually stop feeling your watch on your wrist and aren't bothered by the rims of your glasses in your field of view. Stepping back and taking a second to notice these dismissed details can give you a new appreciation for them.

The earliest known representation of the Moon's features may be a 4,800-year-old Neolithic carving, etched into a rock tunnel in Ireland. But for a long time, researchers thought that Jan van Eyck, the painter, was the first to capture the Moon in a semipermanent way in the late fourteenth and early fifteenth centuries. Before that, others thought Leonardo da Vinci was the first, with a sixteenth-century sketch. In 1610, Galileo made a lovely drawing of the Moon's surface as seen through the first telescope.

You can join the ranks of ancient, medieval, and Renaissance artists in making your own observations and leaving them for posterity, so future folks can rifle through your belongings in 2253.

The easiest features to note are the Moon's "seas": dark, flat places covered over by ancient volcanic eruptions. In these features, East Asian observers saw a rabbit, holding a mortar and pestle. Indigenous American legends of Moon rabbits reflect a similar interpretation.

Others have seen, in the Moon's dark and light areas—the light ones being highlands—a face, a man carrying sticks, or a woman and her children. The craters, some of which are substantial enough to see with the naked eye, have led modern people to (mostly jokingly) wonder if the Moon is made of cheese.

These differing interpretations speak to the human mind's desire to see patterns where there is randomness, and to, in particular, shape those patterns into familiar terrestrial objects. That's inaccurate, of course—there are no rabbits, men, women, or faces in the Moon's magnificent desolation. But imagining that there are can be (a) fun and (b) a good way to notice what normally passes by unperceived. You may also notice, when you start to look closely, that the patterns of light and dark do not change. The Moon is rotating, like most objects in space, but it is orbiting the Earth at the same pace, so we always get nearly the same view, the same face staring back at us.

In this exercise, you'll make a portrait of the Moon—noticing its specifics like you would the freckles on the face of a friend—and, if you're up for it, adopt some new patterns for yourself.

WHAT YOU'LL NEED

This book, a notebook, or a sketchbook
A pencil or pen

WHAT YOU'LL DO

You'll go check out Earth's closest neighbor and take a few minutes to observe the distinctive features that most of us ignore all the time, just because—like the framed family photo on the wall—we look at it so often that we have stopped seeing it.

THE DETAILS

- Using a lunar calendar—on an app, on a desktop calendar, or from an internet search—figure out when the next Full Moon will occur.

- A few hours after sunset, when the Moon is high enough in the eastern sky that you can see it unobscured, go outside and find a spot nearby where its disc is fully visible.

- While the Sun would irreparably scar your retinas, you can look directly at the Moon—no problem. Have at it!

- In the following space provided or on your own paper, draw what you see. Take careful note of darker and lighter regions as well as places that seem higher or lower in elevation.

- See if these different features look like familiar objects to you, and make your own Moon mythology if you feel like it.

- Remember the whole time: you're looking at an actual body in space, spinning like a top as it moves around the planet on which you live. And—bonus—humans once walked around up there!

KEEPING IT SHORT

Just go look at the Full Moon whenever it next appears and in whatever amount of time you have. Take note of the features and areas of dark and light that you can see with the naked eye. Do those areas look like anything to you?

THINGS TO PONDER

- Have you ever looked at the Moon closely before? When was the last time you remember staring at the Moon, and what were the circumstances?

- If the Moon were to suddenly and inexplicably disappear, how long do you think it would take you to notice? Would you miss it?

- The craters and seas on the Moon's surface tell us that it is a very old surface, formed billions of years ago. Observing it closely is a way to look back in time. Are there things you encounter on Earth regularly that give you a look back in time, even if they're just initials etched into a sidewalk?

- Have you ever seen recognizable shapes in the light and dark areas on the Moon? What do you see?

MINDFULNESS EXERCISE (VISUALIZATION)

Picture yourself on the surface of the Moon, feeling light from the lack of earthly gravity. Imagine yourself bouncing from crater to crater, scaling a large mountain with ease (provided you have the right imaginary breathing equipment), and moving from light to dark regions. Each step launches you high into the air, and you land lightly. The Moon in your mind won't be the Moon as it truly is—unless you have an amazing mental map and intellectual encyclopedia—but that's not the point. Just explore the version of Earth's satellite that your brain creates, taking note of whatever details it may dream up, and exploring them for however long you have. It's like freewriting but for lunar exploration.

OBSERVING A LUNAR ECLIPSE

Has your cat ever stood between you and the TV? It turns out that lunar and solar eclipses aren't really that different from this common domestic situation. These celestial events also happen when one thing gets in the way of something else. A lunar eclipse occurs when the Earth comes in between the Sun and the Moon, the Moon moving into Earth's shadow.

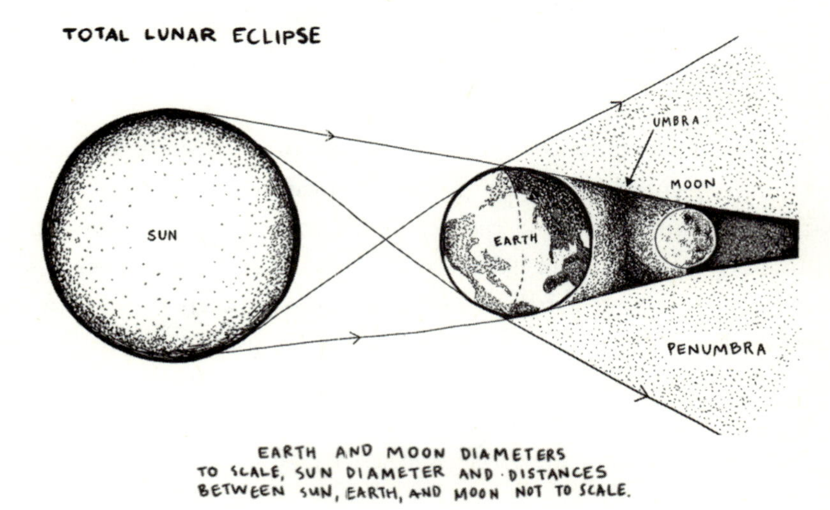

A solar eclipse happens when the Moon comes in between the Sun and the Earth, with the Moon blocking the Sun either partially or fully from us on Earth.

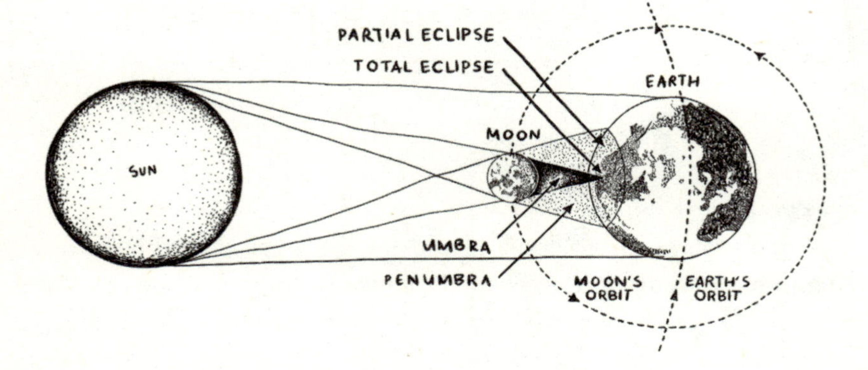

EARTH AND MOON DIAMETERS TO SCALE, SUN DIAMETER AND DISTANCES BETWEEN SUN, EARTH, AND MOON NOT TO SCALE.

Lunar and solar eclipses are some of the most moving of celestial events. Watching a familiar object in the sky change slowly into something you don't recognize can feel both beautiful and frightening. Humans have long observed the patterns of lunar and solar eclipses, and even in ancient times predicted them quite accurately. Eclipses were and are awe-inspiring events. Lunar eclipses, which will only happen when the Moon is in its Full phase, are more common, and you can see most of them over much of Earth's surface.

Solar eclipses are rarer, and you can see them only from limited "bands" on the Earth's surface. As mentioned previously, the next total solar eclipse will not be visible in North America until April 8, 2024. Between now and then, you could observe three total lunar eclipses from anywhere in the US. In this exercise, you'll focus on observing a total lunar eclipse.

WHAT YOU'LL NEED

Open eyes and an open mind
This book, a notebook, or a sketchbook
A pencil or pen
Art supplies (optional)

WHAT YOU'LL DO

Unlike many of the exercises in this book, this one begs a little planning. Lunar eclipses don't happen every month! Instead, you'll have to pick out an upcoming lunar eclipse and then, at the appointed time, go inhabit the experience. Eclipses are visual events, for sure, but you may experience the moment with your other senses as well. While this exercise does not ask you to make specific observations or drawings, there is space to react to this event in any way that feels right to you.

THE DETAILS

- Search online for "lunar eclipses near me" to determine exactly when this type of celestial coincidence will next occur. Here is a partial list:

UPCOMING TOTAL LUNAR ECLIPSES VISIBLE IN NORTH AMERICA

May 16, 2022
November 8, 2022
March 14, 2025
March 3, 2026
December 31, 2028 (western US)
June 26, 2029
December 20–21, 2029

- Pick an event, whether it's a partial, penumbral, or total lunar eclipse. On a website like timeanddate.com, you should be able to find more detail about the event, such as, for a penumbral eclipse, the time when the Moon will fall into the outer part of Earth's shadow. With a full eclipse, it will list when the Moon will fall into the deepest part of the Earth's shadow.

- Set a reminder on your calendar for this observation!

- When that date arrives, find a comfortable spot from which to watch the eclipse's progress. It may be from a back porch or patio, or from an apartment window. Or you might be able to nab a patch of grass in a park or abandoned baseball field nearby. Wherever you go, try to find a place where you can focus your attention on the event.

- Make a note of your start time.

- As the Earth's shadow begins to cover the Moon, what do you notice? How does its brightness change? How does its color change?

- At the peak of the lunar eclipse, when the Moon is in the deepest part of the Earth's shadow, what do you notice? How do the brightness and the color of the Moon change?

- Make a note of the time when you end your observations.

- Use the space provided to write or draw your reflections about this event.

KEEPING IT SHORT

Look online to find the time of a maximum lunar eclipse in your location. This will be when the Full Moon is in the deepest part of the Earth's shadow. Go outside close to this time, and observe the Moon for a few minutes, trying to focus on its shape and its color. Use the space provided in this exercise to write or draw something that arises from this experience.

THINGS TO PONDER

- Have you ever watched a lunar eclipse before doing this? How did that event make you feel? Is it something you would want to experience again?

- ⊙ If you didn't know what was going on up in the sky, how might you interpret a lunar eclipse? Would you be afraid of it?

- ⊙ Eclipses are rare but predictable events that repeat over the eons. What other natural cycles are like that? And what do foreseeable-but-still-surprising things add to our experience of life on Earth?

MINDFULNESS EXERCISE (CAREFUL OBSERVATION)

During a lunar eclipse, one of the main changes in the Moon is its color, and noting the changes in its brightness and color during an eclipse can be a moving experience. Wherever you find yourself, take a full minute to notice the colors around you. Are they bright? Subtle? Is there one dominant color? Notice how the colors make you feel in the moment.

WHY DOES THE MOON CHANGE?

The Moon constantly changes its phase and position in the sky, waxing and waning and shifting over the course of each 29.5-day cycle. Every lunar cycle begins with a New (dark) Moon, slowly slides up in brightness and visibility to a Full Moon, and then slips back toward darkness and "invisibility" as it approaches the New phase again.

The phases between each New Moon—when the illuminated half of the Moon faces away from the Earth—follow a regular pattern that repeats cycle after cycle:

1. New Moon
2. Waxing Crescent
3. First Quarter
4. Waxing Gibbous
5. Full Moon
6. Waning Gibbous
7. Third Quarter
8. Waning Crescent

People used to link the Moon's phases to everything from fertility to crime rate to mood. No scientific analysis has ever confirmed these connections. But that doesn't mean Earth—or its inhabitants—are immune to the Moon's effects. The latest theories suggest this part of this cratered satellite actually came *from* Earth: another planet-size object slammed into our home in its very early years, and the

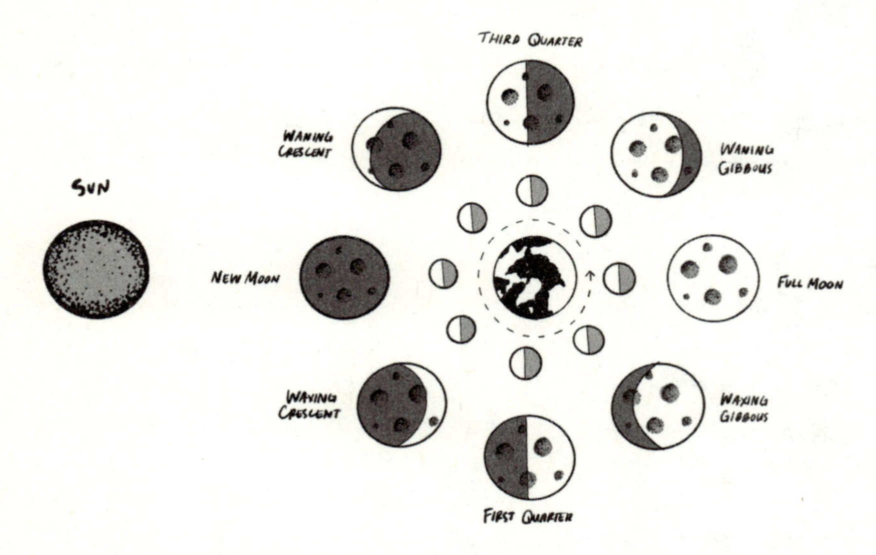

The portion of the always-illuminated half of the Moon that we can see from Earth changes every day.

debris of this catastrophic collision coalesced into the Moon. That accident—violent though it must have been—is part of the reason humans exist at all. The incoming planet may have delivered some of the molecules necessary to make life. And once the Moon was in orbit, it gradually slowed Earth's rotation. The Moon tugs on Earth's oceans to form tides, potentially encouraging life to thrive both on land as well as in water, as it created zones that were sometimes wet and sometimes dry. Its gravity may even affect the planet's plate tectonics, tugging on land just as it tugs on the oceans, and changing the stresses on the planet as the tides go in and out.

So while the Moon probably won't make you mad—a *luna*tic, as people used to say—it did in a lot of ways help to *make* you. Paying some attention to that reflective circle in the sky, rather than the glowing rectangle in your hand, can help you regularly consider all the chance incidents, twists, and turns that led to life on Earth, a reflective line of thought that you can apply to your personal, present life. What events, especially ones that seemed negative at the time, actually made you the person you are at this moment? What are the colossal impacts you survived that made you who you are today?

This exercise gives you the mental and chronological space to contemplate your formation and change, as you create an artistic rendering of the Moon's progression through its phases. All you have to do is note the time and your location whenever you're able to go out, look up the Moon, and contemplate. If you miss days, that's fine! The Moon will still be there when you get back.

WHAT YOU'LL NEED

This book, a notebook, or a sketchbook
A pencil or pen

WHAT YOU'LL DO

For however many days as you are willing and able during a month's time, you'll go outside around the same time each day, take note of where the Moon is, draw it, and chronicle the changes in its shape and position over the course of a full cycle. In looking at all the data you'll collect, you should see the Moon change from New to Full, and every state in between. You'll learn how these changes, and the shift in when you can see our neighbor at all, reveal where the Moon, the Earth, and the Sun are relative to one another. As you do this, hopefully you'll consider the series of events that made this eternally orbiting Earth-Moon pair, and how they made you.

THE DETAILS

- Find out when the Moon will set on the day you'd like to start your observations. For this, you can use the moonrise/moonset calculator at timeanddate.com/moon or an app (see appendix B, page 193).

- On that day, go outside sometime close to but before the Moon dips below the horizon, pencil and notebook or sketchbook in hand. Note that the Moon will set at a decent hour when it's in a Waxing Crescent phase. This might be a good phase to start in.

- When you find the Moon (it shouldn't be hard—it's the biggest, brightest thing up there), draw it as carefully as you can.

- Write a note about how high it is above the horizon (or you can draw a horizon on your page and then draw the Moon above it), and what cardinal direction it's in (north, south, east, west). You can use a compass or download a compass app for this part.

- Also write a note about what the weather is like and any other observations that strike you about the Moon itself or the conditions around you and it. All this requires paying attention to your present surroundings and placing yourself relative to them. Take a moment to consider your position in space.

- Repeat this process as often as you're able during a thirty-day period, making notes if a change between observations seems significant. Also jot down personal or celestial thoughts—about cycles and shifts—as a kind of journal. Make sure to go out around the same time each day until you can't see the Moon anymore at that hour.

- After you can't see it anymore, estimate what time you'd be able to see it the next night and give it a shot! It's okay if you don't see it—record that in the observations too, and try again another night.

KEEPING IT SHORT

Wait until you notice the Moon at night. It can be in any phase and at any position. Just note the time and the shape of the Moon (do a quick drawing if you can). On any successive clear night (it can be a day, a few days, or a week later), look for the Moon. How is its shape different? How is the position different? Jot down any personal or celestial thoughts you have—about cycles and shifts—as a kind of journal.

THINGS TO PONDER

- ⊙ How often, if at all, do you notice the Moon in the sky during the day?
- ⊙ Every change in the Moon affects the environment around it. How does the Moon's presence at night influence what the rest of the sky looks like?
- ⊙ If you have access to a telescope or binoculars, use one of them to look at the Moon during a smaller (Crescent) phase. How is the experience different from looking with just your eyes?

MINDFULNESS EXERCISE (CAREFUL OBSERVATION)

One goal of this observation is to focus on small changes, night after night, to the size and location of the Moon. As you draw, try to clear your mind of what the Moon "should" look like, and focus on what you're seeing each night. Do you notice mottled areas of white and gray on its surface? Try to draw your view as carefully and deliberately as you can. You can practice this skill by doing a "blind contour drawing" of a piece of fruit or a flower. Without looking at the paper, try to capture the outline and details of the object by moving the pencil on the paper as you move your eyes around the edge of the object you're drawing, connecting what you see with your pencil as it touches the paper. This type of drawing should help you to practice observing any object (including the Moon) in its full detail and connect you to the object you're drawing.

THE UPS AND DOWNS OF THE OCEAN

Have you ever been sitting or lying on a beach, basking peacefully in the Sun, when a wave washed over your feet and you realized, *I have to move or I'll be in the surf soon.* Well, if so, now you have something to blame: the Moon. Most scientists agree that humans wouldn't be here if the Moon weren't. This pockmarked satellite has slowed Earth's spin, giving the globe reasonable weather, since the planet's rotation drags the atmosphere with it—which dictates the climate. It also stabilizes the Earth, so it isn't flipping over all the time. The Moon's orbital path keeps ocean levels higher at the equator than at the poles, which are less friendly to biology, giving more real estate to sea life.

That mixture of stability and dynamism helped make Earth a friendly place to foster the little organisms that were the first to swim around in the primordial soup. Today, we're more used to hearing about silly ways our lunar neighbor might affect life and humans here—causing an uptick in crime, causing craziness, causing werewolves to emerge from a dark forest. The true lunar influences are harder to find, but they exist: Lions eat less when it's light at night and hunt more during the daytime after a Full Moon, maybe making them hangry, with all the additional light causing hunting to be more difficult. Corals undergo mass spawning (sexy) based on the strength and timing of moonlight. Scorpions turn blue under the lunar glow and, maybe because they become so flamingly visible during a Full Moon, mostly hunt during a New Moon.

But none of that is the gray satellite's most powerful influence on biology: all this life might not look the way it does today without the ocean's tides.

As the water flows out on terrestrial coastlines, at low tide, it reveals formerly submerged land. Hours after that, the land goes back under the lapping waves. Scientists think this—pieces of the planet that are both dry land and not—forced life to adapt, becoming amphibious and then land-dwelling. Scientists also think it might have catalyzed the chemistry that helped life arise in the first place.

In this exercise, you'll watch the ways that the rushing and receding of salty waters coincides with the phases of the Moon—charting the same forces that pushed and pulled on our most ancient ancestors.

WHAT YOU'LL NEED

A notebook or a sketchbook
A pencil or pen
A tide table, a tide-telling app (like Tide Charts or My Tide Times), or NOAA's Tide Predictions, which you can find by searching online
A Moon calendar or app, like the aptly named Moon Phase Calendar

WHAT YOU'LL DO

Picking either your location or a favorite coastal spot anywhere in the world, you'll look up the first high and low tides during different phases of the Moon and determine the difference between them. By mapping out the gravitational forces acting on Earth, you'll determine why the Moon has the effect it does on the planet's saline covering and investigate the invisible connections between this lunar world, our world, and the Sun.

THE DETAILS

- Using a Moon calendar of your choosing, note when the First Quarter and Full Moon phases will be happening.

- Write down a day for each in the following table.

LUNAR PHASE	DATE	LOW TIDE HEIGHT	HIGH TIDE HEIGHT	TIDE DIFFERENCE

- Pick your favorite seaside location on Earth. Take a moment to think about what that place is like and why you like it. White sands? Cold, rocky shore? A sweet break? Forever views? When will you see that spot again?

- Using your chosen app, paper guide, or website, look up the first listed high and low tide heights for that beach during your two days. Record them in the table too.

- Subtract the low from the high, and record the difference. Which difference is bigger? Consider why that might be, and imagine the water rising and falling, with the relevant Moon phase in the background.

- Now, using whatever feels fun (pencil, colored pencil, crayon, watercolor paint), draw the ocean over a circular Earth in your notebook or sketchbook, imagining what the Moon's influence looks like—with the ocean stretched in the direction that seems like high tide and squished in the direction that seems like low tide. As you look at this picture, or the one provided below, imagine that the Earth revolves on its axis every day, but the Moon goes

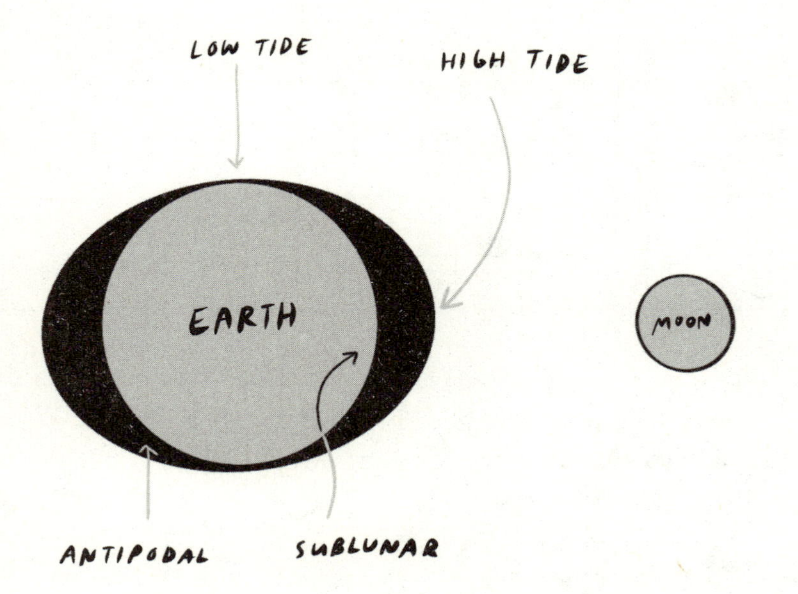

An exaggerated version of the Earth's tides.

around only once a month. Think about why this means that you will have two high tides (and two low tides) every day.

KEEPING IT SHORT

Draw what you think the Earth, Moon, and Sun look like relative to one another during the highest high tide and the lowest low tide, at the spot on Earth where you are.

THINGS TO PONDER

- Are there instances in your life when the timing of tides has mattered to you?
- Would you feel any differently about the oceans if there were no tides?
- How, if at all, would life be different if the Moon and Sun were never up at the same time? What about if they were always up at the same time?
- Do you notice the tides when you're at the beach? How do the changing tides affect your experience of the beach?

MINDFULNESS EXERCISE (PRESENT MOMENT)

Find the nearest body of water: a nearby lake, swimming pool, or beach—or if that's too far, your bathtub. Dip your feet in or go in all the way, depending on how wild you're feeling. Beginning with your toes, notice the way the water feels against your skin. Feel it pressing against your body. Notice its slight undulations. Each time you feel one of those little waves, picture the Moon far above you and imagine that its gravitational force is pushing and pulling on that water, pushing it against you and pulling it back. Experience this moment in all its detail.

INTERLUDE:
AUSTRALIAN ABORIGINAL ASTRONOMY

The observer and the observed are the same entity.

RAGBIR BHATHAL, AUSTRALIAN ASTRONOMER AND AUTHOR, A LECTURER AT THE WESTERN SYDNEY UNIVERSITY

Astronomy classes today typically start with a review of the past few thousand years. In most textbooks, cosmic study begins with "ancient" astronomy as practiced by Babylonian and Greek scientists, and then jumps to the sixteenth and seventeenth centuries, portraying the rapid innovations in technology and theory that ushered in modern Western astronomy.

This sort of review misses the knowledge and understanding that stretch back tens of thousands of years. Take the Indigenous peoples of Australia, a continent that represented a jumping-off point for early travel to Polynesia. Those journeys were propelled by a detailed understanding of the motions of the Sun, Moon, stars, and planets that helped brave travelers navigate both space and time.

Ragbir Bhathal is an Australian astronomer and author, and a lecturer at the Western Sydney University. He recently pointed out the deep differences between modern-day, non-Native astronomy and astronomy as practiced by Native Australians:

> In Aboriginal astronomy the origin of the universe goes back to a time called the Dreaming. . . . The Dreaming is not only an ancient era of creation but continues even today in the spiritual lives of the Aboriginal people. All life—human, animal, bird or fish—is part of an ever-transforming system that can be traced back to the Spirit Ancestors who go about the Earth in an eternal time called the Dreaming. As these spirit people roamed the Earth they made the mountains, rivers, the sky with its celestial objects and all the other features we see in the natural environment around us. The Aborigines are in fact co-creators of the universe they live in. The observer and the observed are the same entity.[1]

Over a hundred years ago, activist and author James Dawson noted that "of such importance is a knowledge of the stars to the Aborigines in their night journeys and of their positions denoting particular seasons of the year, that astronomy is considered one of the principal branches of education."[2]

In Indigenous Australian astronomy, many stars and groups of stars sync up with societal rules. The red star Aldebaran (Karambal), for example, represents infidelity. Its red color and brightness refer to the flames that consumed the adulterer Karambal. For many Aboriginal tribes, the appearance of certain constellations portends particular food harvests and animal behaviors. When the cluster of stars known as the Pleiades first shows up in the dawn sky in the fall, one tribe knows that the dingo breeding season is beginning. These connections between the astronomical (Pleiades) and the practical (animal breeding) are widespread in Australian Aboriginal practices.

Australian tribes grouped stars differently from Western astronomers in some cases and similarly in others. Interestingly, Orion (Seg) and the Pleiades (Usiam) represent a man and a group of women in both traditions. Other Aboriginal tribes see Orion (Seg) as a pack of dingoes chasing the

Pleiades (Usiam), a group of kangaroos. The constellation of Tagai, a sea hero, is very large and includes the western constellations of Scorpius, Lupus, Centaurus, Crux, Corvus, part of Hydra, and one star in Ara.

In Aboriginal traditions, "everything on the land is reflected in the sky," says Dr. Duane Hamacher, an astronomer and senior research fellow at the Monash Indigenous Studies Centre. "The sky is a textbook, it's a law book, it's a science book." Hamacher has spent time with many Australian Aboriginal leaders, talking about the relationship between astronomy and everyday life. Like this: Elders "read the twinkling of stars" and look at precisely how they fluctuate. Rapidly or slowly? Are the stars sharp or fuzzy? They can use that shimmer to predict weather, seasonal change, and trade winds. If the stars look blue and fuzzy, for instance, grab an umbrella.[3]

Science is encoded in this knowledge. On Moa Island, elders use the position of the setting Sun to determine the time of year as well as the cultural and economic activities that should be happening. From one particular location on Moa Island, the Sun sets over one small island on the winter solstice and over another island on the summer solstice. Knowledge is enmeshed in the landscape, particularly in the connection between the local terrain and the sky.

These traditions, says Hamacher, go back many centuries, and they speak to the strength of oral cultures, in which knowledge transcends generations, without degrading. Aboriginal people associate a memory with an object, and the object can be the landscape or the night sky, or some interaction between them: the location of the Sun on a particular day or the rising of a particular constellation.

That, in itself, is mindful. Because mindfulness, in part, is about paying close attention. And as Hamacher puts it, "For tens of thousands of years, Aboriginal people . . . have paid incredibly close attention to the world around them, and still do today."[4]

INTERLUDE:

PROFILE OF ASTRONOMICAL EDUCATORS KARLIE NOON AND WILLY STEVENS

Astronomer Karlie Noon, a Gamilaraay woman who grew up in New South Wales, was working at a science summer camp when she met a researcher in a community she had newly discovered: an astronomer who studied not just the universe's history but also the history of humans' efforts to observe, interpret, and put to use sky knowledge.[5] Soon she began to study Indigenous understanding of a strange lunar phenomenon called "Moon haloes," which occur when ice crystals in Earth's atmosphere change the path of moonlight, causing rings to form around that bright, pocked sphere.

Indigenous peoples had long used these haloes to predict the weather: the sharpness of the structure, and the stars between the Moon and its ring, revealed to observers how much moisture was above their heads, and how quickly they needed to prepare for a storm. With telescopes and satellite instruments, humans now have a harder-data understanding of the atmospheric conditions that cause tempestuousness and Moon haloes. But Indigenous astronomers had, long before that, taken enough naked-eye data to identify patterns and understand what the sky's state meant for life on the ground. Today, Noon is gathering settlers' written stories about Indigenous peoples' lunar-ring observations, to learn what they knew so long ago.

Noon has been part of the organization Australian Indigenous Astronomy, which helps gather and distribute resources and research about the continent's long history of cosmic quests.

Understanding the connection between Indigenous science, history, and culture is also important to Willy Stevens, a Muruwari man and associate at Australian Indigenous Astronomy, who—like many kids on the continent—didn't grow up with knowledge of his cultural scientific history.

Stevens began to encounter more Aboriginal culture when he met his biological mother at age seventeen, years after a Kamilaroi family adopted him. His mother spun him traditional stories—including some about the sky. Later in life, he joined the Sydney Observatory as a public educator and began presenting the "Dreamtime Astronomy" program, where he was able to share with others the kinds of stories his mother told him—and the

Aboriginal astronomical painting.

knowledge that they're connected to and a continuation of their Pacific past, in the same way the Earth is connected to the Big Bang. "I feel very humbled when Aboriginal people thank me after coming on my tour and discovering things about their culture they had never known," he told Australia's *Cosmos* magazine. "Learning these stories gave them a sense of pride and made them want to learn more."[6] For Stevens, that pride and knowledge aren't important because they relate to the cosmos but because they illuminate history, morals, tools, ceremonies, and traditions that might otherwise get lost on the colonized Earth.

IV. THE STARS

IN WHICH YOU EXPLORE ALL THE OTHER SUNS IN THE SKY

HOW MANY STARS CAN I SEE?**

HEARING THE UNIVERSE**

DON'T CRY OVER SPILLED SKY MILK, OR FINDING YOUR HOME GALAXY*

OBSERVING THE BRIGHTNESS AND COLOR OF STARS*

WHY CAN'T I SEE ORION TONIGHT?***

It seems that humans have always looked at the stars, forming the unconnected, scattered points of light into familiar shapes: bears, hunters, twins, gods. These constellations give us a way to impose order and familiarity on strange, distant objects. And in many ancient cultures, the constellations were a way to memorize traditions and even laws. The people who named those constellations didn't know that the stars were, you know, *stars*, like the Sun but much farther away. They didn't know those stars warmed exoplanets. In fact, they didn't even know that our planet orbited our own Sun.

But they did know and notice many things about stars. They watched the constellations rise and fall each night. They knew which ones were visible at what time of year. If they were far enough to the north, they knew how to find the mysterious one in the northern sky around which all the others seemed to turn. And although they didn't know that the glow of the Milky Way encompassed hundreds of billions of stars too small for our eyes to resolve, they saw that glow and told stories about it, whereas most modern citizens (occupying bright cities) have never seen the Milky Way at all.

The skies are polluted with light in many places. The horizon is blocked by buildings. But even without a pristine dark-sky environment, there's a lot of stellar (pun intended) activity to watch up there if you know where to look. And if you can situate yourself among the other stars in our galaxy, it's a little bit easier to imagine—and really feel—like you're connected to those stars, our local solar system being just one of billions upon billions orbiting the galactic center.

HOW MANY STARS CAN I SEE?

Sometimes it's hard to remember that stars speckle the sky at all, let alone to stop and consider how they're roiling spheroids of plasma, fusing their hearts out. But taking stock of how many other suns you can see lends a sense of perspective to the size of the universe and your place in it.

In this quick exercise you'll figure out how many stars you can see in the sky above you. You'll see a different number if you live in the country versus the city, but you want to pick out an area that is as dark as possible, depending on your circumstances. Your backyard might be fine. If you live in a city, maybe go to a park or to the roof of your building.

It would be very difficult, if not impossible, to count all the stars in the sky. Don't let us stop you from trying if you've got some time to kill, but it might be easier to estimate instead. As a cheat, you can count the stars visible in a small patch of sky and then multiply your count by the number of that-size patches it would take to cover the entire sky.

Consider what this process means for the way the universe is, and the way you are in it. Each of those stars beaming into your viewing tube is another sun—and probably another whole planetary system. Even distant, seemingly foreign spaces are really just variations on the theme of *this* solar system, and it's only by coincidence that we find ourselves in this particular one.

Using a few random areas to infer information about the rest of the sky can also help you consider the relative sameness of the universe: stars must be spread pretty evenly (give or take), and the same is actually true of the universe's larger structure, where galaxies are laid out in a predictable pattern. No matter how different one spot (in the universe or on Earth) seems from another, they're actually pretty similar.

WHAT YOU'LL NEED

A clear nighttime sky
This book, a notebook, or a sketchbook
A pencil or pen
An empty toilet paper roll

WHAT YOU'LL DO

You'll hold your toilet paper roll up against your eye and peer through it at random patches of the dark night sky, counting the number of stars you can see in each patch. Then you'll use those counts to determine the number of stars visible to the naked eye in the rest of the sky from your location.

THE DETAILS

- Hold the toilet paper roll up to your eye and look at a piece of sky not too close to the horizon. Yes, you'll look silly. Embrace the cringe!

- Pick other regions randomly. Try to look at a variety of spots around the sky.

- If you don't see any stars, don't shift to another region to recount. Just record zero for the number of stars visible.

- ⊙ Make a total of five star counts through the viewing roll.
- ⊙ Record each of these star counts in the space provided.
- ⊙ Average your five star counts by adding them up and dividing by five.
- ⊙ Multiply this average by 250, the number of tube "patches" needed to cover the sky to get the number of visible stars where you are. (For mathematically inclined readers: check out the note in the back of the book, which will explain why we use the number 250.[1])
- ⊙ Record your number. Was it larger or smaller than you thought it would be? Now you'll be able to casually say at parties, "On a good night, you can see about 1,274 stars 'round these parts."

THINGS TO PONDER

- ⊙ What might be some things that affect the number of stars you can see?
- ⊙ Which of your friends live in darker or brighter places? See if they want to play along and compare notes. Maybe even try the experiment on your own at their house. Does it make you feel any differently, or think about the universe differently?
- ⊙ You did this exercise with just your eyeballs. How might it change if you used a telescope?

- Do you think humans long ago saw more or fewer stars in the night sky? Why would that be? Would your life be any different if you could see more stars?

- Aside from a toilet paper roll, what might make you more aware of the stars in the sky on a daily basis?

- Where have you visited or lived that displayed the most stars at night? How did that make you feel?

MINDFULNESS EXERCISE (CAREFUL OBSERVATION)

Careful observation can help you see very faint stars. It is somewhat counterintuitive, but true, that if you look slightly to the right, left, above, or below a faint star, that star will seem brighter. Find a star in the night sky that you can barely see. Then look just slightly away from it, and you should suddenly see it better. Try to experience the moment of seeing light from a faint distant star in all its wonderfulness.

HEARING THE UNIVERSE

More than 13.77 billion years ago all the matter and energy in the universe was contained in an infinitely dense point—a point that itself was, at the time, the whole universe. Then the universe burst outward, accelerating, for a time, faster than the speed of light. The cosmic ingredients spread out, the universe itself growing as they did so. These ingredients kept diffusing outward, and the universe kept expanding and then, later, accelerating its expansion, till we got the cosmos we have today (and it looks like it will keep zooming outward into the foreseeable future!). In every direction you look, the universe is essentially the same, with the same basic types and density of galaxies. The Big Bang was a highly effective spreader of the cosmic peanut butter across the bread of space-time.

This Big Bang—the starting point in space and time for our universe, and, really, the starting point *of* space and time in our universe—released a ton of energy, more than you can possibly imagine with any degree of accuracy. And the heat left over from those early fractions of a second stays with us today, now spread across the much-larger universe. The universe's current temperature is about −454 degrees Fahrenheit (2.73 Kelvin if you're a scientist)—not so hot, but remember that it's been stretched across space-time and has been cooling down for 13.77 billion years. It takes your coffee only a few minutes to get tepid.

That leftover heat presents itself to astronomers as radiation called the "cosmic microwave background," which shows up strongly as radio waves. But it's not just astronomers who can detect that radiation:

you can too. And all you need is a regular old-fashioned radio.

In this exercise, you'll use a normal household object to make a connection to the universe's birth. Afterward, it won't be easy to forget that this remnant of cosmic genesis exists all around you all the time, even when you can't hear it.

You'll also use a radio to see what other invisible radiation is swelling through your current spot in space-time, and what's causing the radiation. These radio waves aren't dangerous, but it's interesting to imagine them flowing around you constantly, even though your limited sense organs can't pick them up.

The Robert C. Byrd Green Bank Telescope in Green Bank, West Virginia. Human shown for scale.

At radio telescopes, like the one at the Green Bank Observatory in West Virginia, which boasts a telescope about as big as two football fields side by side, human-produced radio waves present a problem. Here, the strength of radio waves from simple electronics can dwarf that of radio waves coming from distant galaxies, supernova remnants, and black holes. That's why Green Bank is in the middle of the National Radio Quiet Zone, where technology that emits radio waves is strictly regulated. There is no cell-phone signal nearby, residents can't have microwaves or Wi-Fi, and your car radio dial will scan and scan and scan without finding a radio station to play. The fact that such puny devices can cause such problems just shows the massive scale of the universe: space objects are so far away that even though they put out radio waves trillions of times stronger than your cell phone, your cell phone can easily drown them out. It also shows how exquisitely sensitive modern radio telescopes are!

WHAT YOU'LL NEED

A portable radio with a tuner
A notebook or piece of paper
A pencil or pen

WHAT YOU'LL DO

You'll use the radio to figure out what's emitting electromagnetic radiation in your house or neighborhood as well as how the remnants of the Big Bang sound to you in your current position in space-time.

THE DETAILS

- Turn the radio on and switch it to FM, to a part of the dial where you hear only static. (If you don't have a portable radio, skip to "Keeping It Short.")

- Turn the volume up as loud as you want!
- Use the tuning dial to find a spot where you can hear only static—make sure you don't hear any music or talking.
- Listen to that static and take some notes on what it sounds like.
- Walk around your house—or outside if you feel like a trek—and hold the radio up to various items (some ideas: a cell phone, a laptop, a sink, an electrical socket, a light fixture). Take notes on whether getting close to these objects changes the sound coming through your radio, in tone, pattern, or volume.

KEEPING IT SHORT

Turn on your car radio, tune it to static, and simply think about the amazing strangeness of what you're listening to: the 13.77-billion-year afterglow of the origin of the universe.

THINGS TO PONDER

- Your cell phone is a powerful radio receiver and transmitter. How do you feel about how many invisible waves are flying through the air all the time?
- How is that feeling different when you can *hear* versions of those waves on the radio?
- Is the sound itself calming? Annoying?

- What might the night sky look like if you had "radio eyes" rather than eyes sensitive to visible light?

- What other "invisible" things are all around you all the time that you may not notice?

- Consider which of the objects you looked at made the loudest (or strangest!) sounds. What, if anything, do they have in common?

- As for the cosmic microwave background, hotter things emit radiation at higher frequencies—the highest temperature astronomical objects emitting gamma rays. Do you think that in the universe's earlier years you would have been able to pick it up with a radio? Why or why not?

MINDFULNESS EXERCISE (CAREFUL OBSERVATION/PERSPECTIVE)

Turn on your radio, and turn the dial till you hit pure static, the hiss sometimes called "white noise." Sitting comfortably or lying down, close your eyes and listen to it. Focus on the sound, letting other thoughts arise and then pass through your mind. See if you hear any subtle variations in the static, imagining that what you're hearing is partly the universe being born—long before you were here.

DON'T CRY OVER SPILLED SKY MILK, OR FINDING YOUR HOME GALAXY

In elementary school, you likely learned your address: house number, street name, city, state or region. But if your education (and the postal service) were not so Earth-biased, it could have asked you to think about where that house and street were in relation to not just other parts of the planet but other parts of the solar system, the galaxy, and the universe.

That orientation is obviously not that useful for getting your latest package delivered to your mailbox or telling a delivery person where to drop your food. But this understanding is good for thinking about where you fit into the wider universe. Were you to expand your address to include this cosmic context, it would place you on Earth, in the solar system, in the Milky Way galaxy, which resides in a collection of other galaxies called the Local Group, which sits within a group of groups called the Virgo Supercluster, which seems to be part of a larger supercluster called Laniakea, which is itself, of course, just a tiny part of the universe.

The Milky Way, probably the most salient (and highly observable) part of this address, is a spiral-shaped galaxy that contains between 100 billion and 400 billion stars. Earth lives in a small connector between two spiral arms, called the Orion Spur. That location positions us in a kind of galactic suburb, not too close to the galaxy's busy center.

The Milky Way, if you could see it from the edge, looks a bit like a fried egg, with a bulgy center surrounded by a thin disc. When you look out at the Milky Way from your location in the solar system (whose position you can see in the following illustration), you're looking through its disc and, if you can see the galactic center, at its bulge.

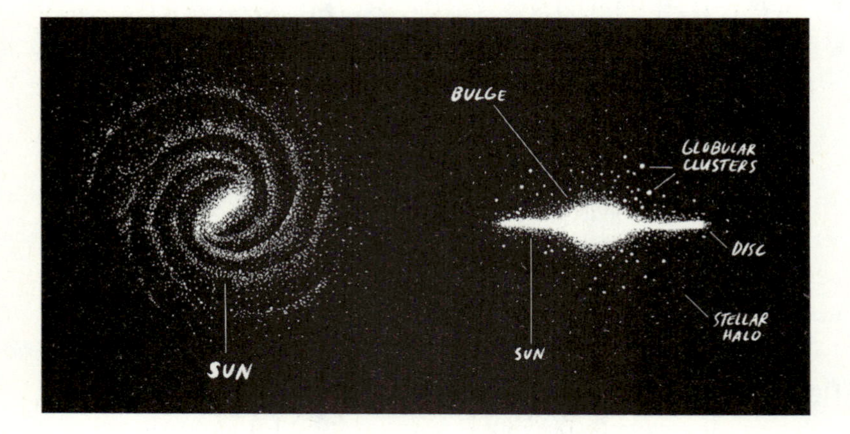

In the sky, the Milky Way will appear as a bright hazy band—almost like a sheen of clouds that doesn't change—stretched across your view of the stars. That's because the Milky Way contains *so many* stars that are so far away, your eyes can't resolve them into individual points but instead see them as a white band spilled above you. Thus the *Milky Way*. In the past, people have envisioned it as spilled cornmeal or seen the dark lanes of dust as the body of an emu.

In this exercise, you'll try to see the Milky Way or at least know in the sky where it should be. That's it!

WHAT YOU'LL NEED

A clear nighttime sky
This book, a notebook, or a sketchbook
A pencil or pen

WHAT YOU'LL DO

Go find the galaxy you live in.

THE DETAILS

- First, you'll need to find a place where you can actually see the Milky Way. That will probably be at least a short drive from where you live, because the lights of a city or town can easily swamp the spilled milk in the sky.

- Maybe you know of such a place offhand. But if not, take a look at the light pollution map found at darksitefinder.com. It shows a color-coded map of light pollution as well as specific locations that have been designated especially dark. If you can, try to reach a yellow-colored zone.

- The Milky Way struts its shine best between June and August, when the bright galactic center is visible, which is a nice time to be outside at night anyway.

- Find a time during those summer months when the Moon isn't too bright—when its phase is close to a New Moon.

- Once you're outside, let your eyes adjust to the darkness for at least ten minutes. After that, this exercise is one of the simplest there is: just look up—no compass, no apps, no star finding. Since the Milky Way stretches across the sky, you should be able to spot it, as an arcing, hazy band. That's your view of where you live—imagine it like you're looking down the hallway through your house or apartment, while you're inside: you don't have the full, overhead view, but you can appreciate it as home.

- If you can't see the Milky Way, use your Star Wheel or an app to find the constellation Cygnus, the swan. The long axis of this constellation—the neck of the swan—runs along the Milky Way. So if it's too bright to see the Milky Way itself, you can use Cygnus to see where it should be visible.

- Take stock of what you notice about your galaxy. Is the milkiness the same brightness throughout the arc? Can you see more individual, bright stars near the band of the Milky Way? If you're feeling artistic, sketch out how it looks, including the scenery as landmarks. You could come back to this same spot every year, or once every five years, and see how things on terra firma and above it have or haven't changed.

KEEPING IT SHORT

If you live in a city and can't get out to a dark spot, just look up some images showing how fancy cameras see the Milky Way. Search for something like "astrophotography + Milky Way." Then next time you happen to be driving through a more rural area at night, you'll know what to look for.

THINGS TO PONDER

- Have you ever been able to see the Milky Way before this? If so, where were you when you saw it?
- If you didn't know stars were stars, or that the haziness was really just a bunch of stars, how might you interpret what you're seeing?
- Where do you visit regularly that might be the best place to go to see the Milky Way?

MINDFULNESS EXERCISE (BREATHING/PRESENT MOMENT)

At night—either at a dark-sky site or just in a park near your house—lie down on the ground with your arms relaxed at your sides and your palms facing upward. Be aware of your breathing, as you inhale and exhale, but do not try to alter it. Look up at the sky and check in on each part of your body, starting with your toes and moving up to your head. Notice how each feels, how your clothes and the ground feel against it, and whether your body is relaxed or tense. If your muscles are contracted, try to relax them, keeping your gaze all the time on the sky above you.

OBSERVING THE BRIGHTNESS AND COLOR OF STARS

As you become more familiar with the night sky, the subtle differences from star to star become more apparent. Instead of a sea of white points of light, all the same intensity, you begin to notice that each star has its own personality of color and brightness. The color reveals temperature, and the brightness mixes together the star's intrinsic shine and its distance from the solar system. (Flip ahead to appendix D, page 201, if you like, to discover more about the science behind star color and brightness.)

While stars may seem to be laid out on a curved canvas above us, all at the same distance, they are, of course, located at a wide variety of distances, with most of the stars visible to the naked eye being relatively close—in our "neighborhood."

If you could live for hundreds of thousands of years, you would be able to witness how dynamic stars really are, moving, varying in brightness, and even changing in color over their lifetimes. As it is, *your* lifetime is over long before a star's. So what each of us sees in the night sky is a snapshot, a freeze-frame of an ever-changing movie.

WHAT YOU'LL NEED

A dark but clear nighttime sky
This book, a notebook, or a sketchbook
A pencil or pen
Binoculars or a telescope (optional)

WHAT YOU'LL DO

To notice the differences in color and brightness, you just have to do one thing: look. In this exercise, you'll observe one of two constellations and note the relative brightness, color, and location of the stars in that constellation. If it's winter, you'll check out Orion, the hunter. And you'll swing over to Cygnus, the swan, if you're in the dog days of summer (which you might note is also a time of year associated with the first rising of the star Sirius in the constellation Canis Major).

THE DETAILS

On the next page you'll find two constellation diagrams, one for Orion and one for Cygnus. Orion is easiest to find in the late fall to winter months (November to February), and Cygnus, in the summer to fall months (June to September). The diagrams will allow you to focus on carefully observing the visible stars in each constellation.

- Locate either Cygnus or Orion. Their exact locations will depend on your latitude, but from North America, the Orion constellation will be in the southern sky, and the Cygnus constellation will be overhead in the early evening. The following images show roughly what the constellation shapes look like. You should be able to find these patterns in the sky as long as you're looking in the right season and after sunset. (You can also use your Star Wheel to locate these constellations, page 187.)

- Try to line up the respective constellation diagram with what you see in the night sky. If you live in a big city, you might not be able

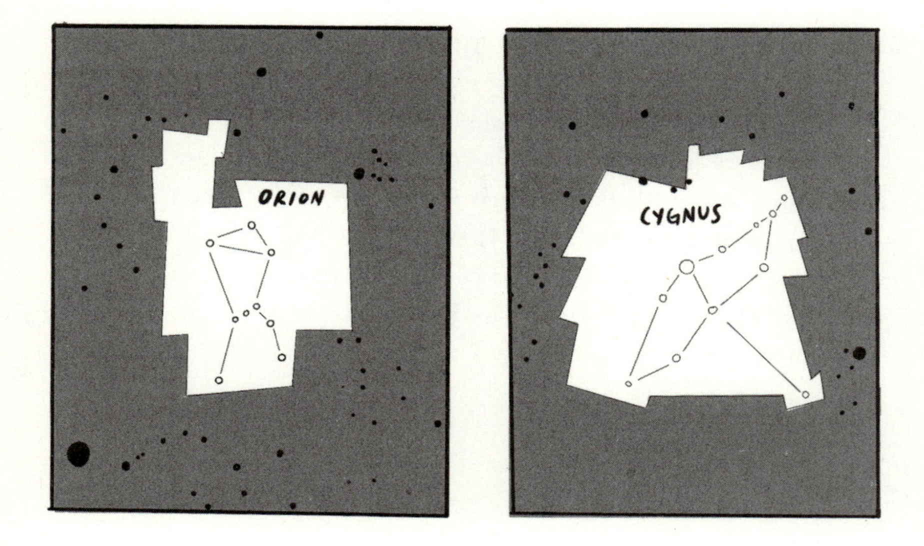

to observe all the stars shown, but you should be able to see most of the bright ones (drawn as larger circles).

- Carefully look at the stars. Do any of them look slightly yellow, blue, or red? Note any colors that you see beside the stars in the diagram.

- Now look at the stars again. Which star is the brightest? Put a "1" next to it. What is the second brightest? Put a "2" next to it. Rank all the stars that you can make out.

- There is extra space on these pages so that you can choose a constellation on your own and do this same exercise with it: make a drawing of the stars you can see in the constellation, make notes about the colors of each, and rank the stars in the constellation from brightest to faintest.

KEEPING IT SHORT

Look at the stars in the night sky wherever you are, any clear evening after sunset. Carefully notice how their brightness varies. Are some clear and bright and others barely visible? Find the brightest star peeping out. Does it have a distinctive color? What is the color? Do you recognize any of the constellations around it?

THINGS TO PONDER

- Recalling that cool stars are redder and hot stars are bluer, did you see more red or more blue stars? Why do you think this might be so? (Hint: blue/hot stars are intrinsically brighter than red/cool stars.)

- A star's lifetime depends on its color. Blue/hot stars have shorter lifetimes than red/cool stars. Which stars would you search near if you wanted to find other life in the universe?

- The stars in Orion represent a hunter in many societies and oral traditions. Does this make sense to you?

- Had you noticed the colors of stars before? How does noticing the colors make you feel differently about the night sky?

MINDFULNESS EXERCISE (CAREFUL OBSERVATION)

Find a comfortable place to sit, inside or outside. Look around yourself and carefully notice the colors that you see—the green leaves of a tree, the color of a wall. Is there a predominant color around you or many different colors? Notice all the shades of a particular color that are visible to you from where you sit.

WHY CAN'T I SEE ORION TONIGHT?

Many of us live in cities with lights so bright that we can see only a few select stars. And when we *do* see stars, we don't often notice which ones are in the sky at a particular time of year. Common constellations, like Orion, are easiest to find, but no matter how carefully you look, you'll be hard-pressed to spot Orion in the summertime.

That's because Earth's trip around the Sun changes which stars are

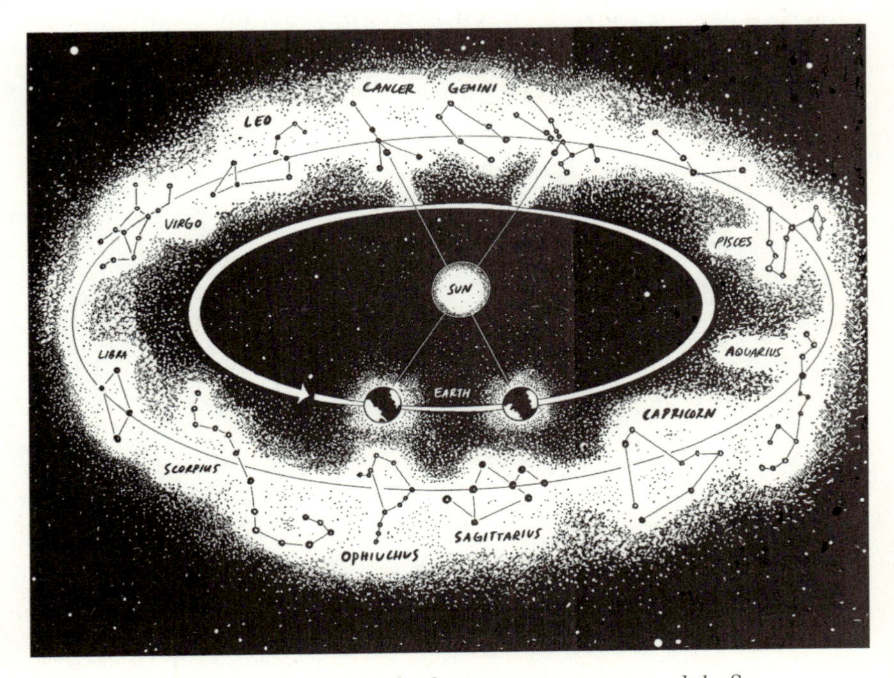

The stars we can see at night change as we move around the Sun.

visible at night. When the Sun is in the same part of the sky as certain stars, we can't see those stars: it's daytime! It can be easy to forget that Earth is a planet moving through space, not separate from the solar system or the stars in the sky.

On planet Earth, we orbit the Sun once every year. As we make this passage, the stars visible near the Sun change. And after a year, the Sun appears to return to where it began, nestled among the same shiny dots. The path the Sun appears to take through the stars, as you now know, is called the ecliptic. And the Sun's apparent path is a circle that surrounds us.

The Sun will seem to pass through a complete circle in the sky once every 365.25 days. Let's do some rounding and just say 360 days for simplicity's sake. (Astronomers do this sort of back-of-the-envelope approximation all the time. It keeps the math simpler!) That means the Sun appears to move about 1 degree, from west to east, each day. That's 7 degrees in a seven-day week and about 30 degrees in an average month. Even on days you're not doing this exercise—stop and consider that 1 degree of motion and our slow, annual progress around the closest star. It's a different way of marking the passage of time, one that's not *quite* so Earth centric.

This exercise will help you get in touch with this idea and note the passing of the seasons. Psychologists have found that marking time in different ways makes time's passage seem slower, in a good way, because the days and years don't simply blend together.

WHAT YOU'LL NEED

A clear nighttime sky
This book, a notebook, or a sketchbook
A pencil or pen
Your Star Wheel

Access to the internet or a cell phone with a night-sky app on it (like Star Walk 2) will help you to find constellations, but it isn't essential.

WHAT YOU'LL DO

You'll figure out where in the sky the constellations and the Sun live. Then you'll see how the Sun moves through the sky, how fast, and why that changes the sky throughout the year. This exercise will give you the tools to notice, relative to the sky itself, what "a year" means.

THE DETAILS

- Look back at the diagram showing how you can use your hand as a ruler to measure angular size or distance (page 57).
- Now go outside around sunset (your smart watch or the internet can tell you what time the Sun will set where you are), and try to find a constellation that is just above the southwest horizon. *This location is important!* (If you've never found a constellation before, check out your Star Wheel, in appendix A, page 187.)
- Even if you don't know the constellation you've located, draw the pattern of the stars that you see in the space provided. Voilà! You've made your own constellation. Try to identify this pattern from night to night.
- Using your hand at arm's length, see how far the stars in your constellation are from the setting Sun. For example, if your stars

are four closed-hand widths away from the Sun's setting spot, that's about 40 degrees.

- Now, using your measurement, make a prediction! Determine how long it will be until you can't see your constellation anymore—when the Sun, appearing to move 1 degree per day, will bump into this collection of stars and outshine its light. In this way, you're seeing the universe the way scientists do: as a space ruled by laws that let you—when you understand those laws—look into the future.

- Test your prediction. If you think your constellation will vanish in a month, go back out in a month and try to find it. You can mark time in this way whenever you want, for however long you want, even without taking notes. If you notice a particular constellation one day, just try to find it on a nighttime walk the next day.

KEEPING IT SHORT

Find or make up a constellation near the southeastern horizon. Make a note on your calendar to come back at about the same time in a month and try to find it again. How far has it moved in that time, if you used your hand to estimate? Pick an upcoming holiday (Thanksgiving, your birthday, etc.), and see if you can find the constellation then.

THINGS TO PONDER

- Why might it have been important for ancient civilizations to notice this annual motion of the Sun through the stars? Why has it become so much less significant to us now?

- The night sky has been a repository for human stories and mythology for millennia. Why do you think this is the case?

- Look at the image of the constellations "behind" the Sun on page 135. Where have you seen all those names before? (Hint: you're such a Leo!)

MINDFULNESS EXERCISE (BREATHING)

Focusing on the night sky with intentionality can provide escape from the persistent buzz of stress in daily life. Try to focus your mind by doing a simple breathing exercise, which will take just a few minutes. Slowly breathe in through your nose and out through your mouth for a minute or so. Imagine the gradually expanding spheres of air reaching out to the stars above you.

INTERLUDE:
POLYNESIAN CELESTIAL NAVIGATION

In the space above, heaven is held fast.
In the space below, held fast is muddy earth.
From the space of heaven to the space of earth, there is still space.

—KEPELINO, NATIVE HAWAIIAN CULTURAL HISTORIAN[2]

Thousands of years ago Polynesians were already navigating the Pacific Ocean. Sailors wended their way from New Zealand to the Hawaiian Islands to Easter Island, and perhaps on to South America—without the benefit of GPS, compasses, or even maps. What they did have, that most of us today don't, was a deep knowledge of the night sky and of how it changed as they sliced across the ocean, from winter to summer and back again.

Some of these skills recently showed up in an unlikely place: a Disney movie. *Moana*'s creators took pains to show how the Polynesian characters would have charted their courses. These skills, and an awareness of the shifting sky, allowed a seafaring society to travel across the open ocean. With this achievement, the ancient Polynesians became the astronauts of their age.

THE LOCAL SKY

Polynesian celestial navigation starts with what experts call a star compass, an example of which follows. This system divides the sky up into thirty-two directions. Here are the ones we're familiar with:

Akau, for north
Hema, for south
Hikina, for east
Komohana, for west

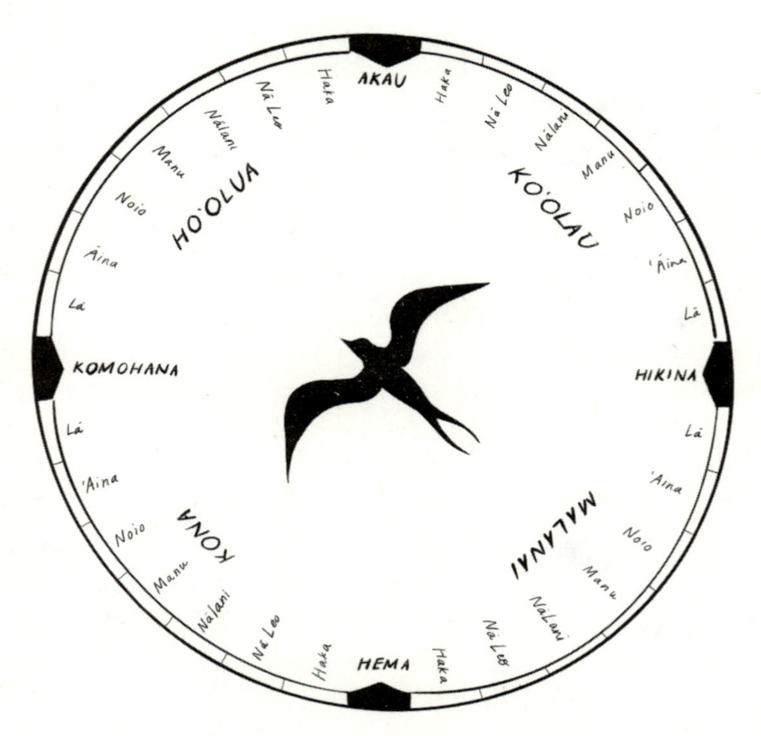

Compass directions all have a unique name in Polynesian wayfinding.

The Polynesian system helps a boat set a bearing depending on what star, constellation, or other object sits at a particular spot on the horizon. Sailors needed to know where those objects would rise and set, depending on the time of year and the boat's location in the vast Pacific Ocean.

STEERING WITH THE SUN

Let's look at a particular example, using the Sun. If you have done the "Position of the Setting/Rising Sun" exercise (page 55), you know that the Sun's position depends on the time of year. On certain days—March 21 and September 21—the Sun will rise due east and set due west. When it's winter in the Northern Hemisphere, the Sun will rise and set a bit south of there; in the summer, it rises and sets a bit north.

Now imagine that—for whatever reason—you find yourself on the open ocean on March 21. You want to head straight north. Assuming the sky is clear and blue, you would keep the Sun directly off the starboard beam at sunrise and directly off the port beam at sunset. When it's light out, you would know that you're skimming along the water due north.

If you find yourself on the high seas later in the year, you know the Sun will rise and set some distance to the north of east and west. You'd just need to keep the Sun lined up with a slightly different marking on the boat.

Of course, the Sun sets, and pesky clouds sometimes obscure the sky. And sometimes you're sailing at night. At those times, Polynesian navigators used the stars to guide their journeys across the oceans.

STEERING WITH THE STARS

To steer a boat with the stars, you just need to know where those stars are and roughly where you are.

For an extreme example, imagine (unlike any Polynesian sailors we know of) you're hanging out at the North Pole. Here the North Star, which these travelers called Hokupa'a, is directly overhead. As you move to the south, the distance of the North Star from the northern horizon is the same as your latitude. When you're at 85 degrees north latitude, the North Star is 85 degrees from the northern horizon. At Hawaii's latitude of 21 degrees north, Hokupa'a chills 21 degrees above the northern horizon.

If a traveler wanted to head north, they would simply keep Hokupa'a directly off the bow of the boat. If they wanted to go south, they would make sure that same star stayed off the stern. Moving in other directions involved keeping particular known stars lined up with the bow or stern of the boat, or with other markings on the boat itself.

Armed with an intergenerational familiarity with the night sky, an understanding of their latitude, and a knowledge of the time of year, Polynesian sailors could move between the islands scattered across the Pacific, day or night.

OTHER SKILLS

While this book focuses on astronomy, Polynesian voyagers used more than the cosmos to skim from island to island. They checked out the behavior of ocean swells, the flight patterns and habits of certain birds, the clouds, and the types of fish more likely to be found close to land. This additional information enriched and deepened the astronomical directions.

RECENT WAYFARING

Since the 1970s, a dedicated group of sailors called the Polynesian Voyaging Society has worked to relearn the skills that allowed ancient Polynesians to discover and settle the Pacific islands. In 1975, they first launched a traditional double-hulled Polynesian sailing canoe named *Hōkūle'a*, the Hawaiian name for the star Arcturus. Not long after, they sailed from Hawaii to Tahiti using only traditional navigation techniques. The *Hōkūle'a*, from 2013 to 2017, traveled around the world. If you want to learn more about this incredible journey, visit the Polynesian Voyaging Society online (hokulea.com/moananuiakea).

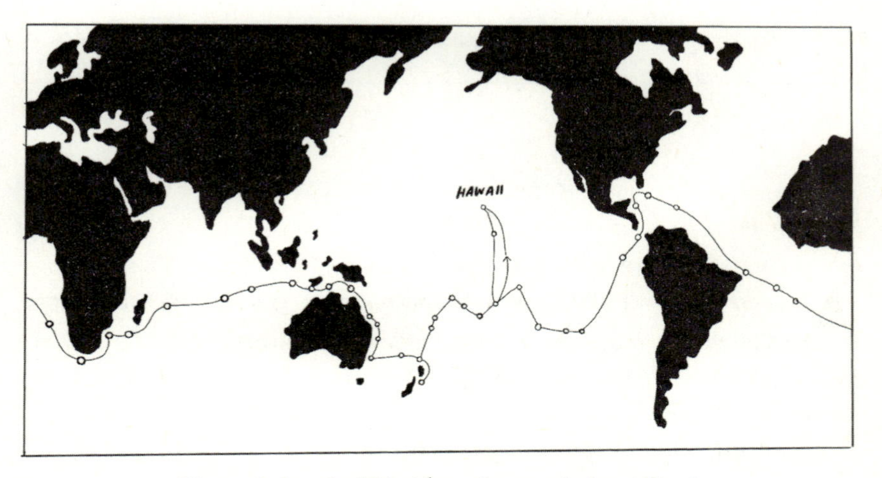

The path that the Hōkuleʻa took around planet Earth.

INTERLUDE:

PROFILE OF CELESTIAL WAYFINDER KĀLEPA BAYBAYAN

"As teenagers, we always heard stories about how our ancestors were great explorers," said Kālepa Baybayan, former navigator in residence at ʻImiloa Astronomy Center of Hawaii, "and they settled these islands by building these large sailing canoes and exploring large expanses of the ocean." He loved to imagine them.[3]

As Baybayan, who passed away in 2021, was graduating from high school, a group of Hawaiians began building *Hōkūleʻa*, a re-created version of those canoes. In 1975, they sailed it to Tahiti, tracing the voyages of ancestral canoes. Seeing it in person, he said, "I was able to add color to my dreams."

When *Hōkūleʻa* launched on its second voyage to Tahiti, in 1980, Baybayan joined as the youngest crew member. "At night, when I would lie on the deck and look up at the silhouette of the traditional 'crab claw' sail cutting an image across the star field, I felt extremely exhilarated," he said. "I understood then that the feeling of excitement and wonder was the same as my ancestors'. I understood I was no different from them."

Soon he became interested in how those charting the course—past and present—did so: How did they determine how to get from where they were to where they wanted to be? Partly by using the stars to navigate. "You use the daytime star during daytime hours, and you use the stars at night," he said. "They are your beacons for setting up your system of guidance. Our non-instrument navigation system is coincidental and unique in that we sail in an area known as the tropics. We're using the equatorial star field to set up our guidance system. The stars rise and set pretty much vertical in their travels from eastern horizon to western horizon. Astronomy allows us to understand how the stars move between the horizons and how we position stars along the horizon to fit into our star compass."

The stars, here, exist in the present—not telling about the past. "We're not concerned about the cosmology of the universe," said Baybayan. "[It's] not necessary for us to appreciate astronomy. But we do have a kindred spirit with astronomers in that we're voyagers, and astronomers are basically explorers."

But hard-nosed navigation isn't the only aspect of travel. "Wayfinding is an Indigenous system of orientation, guidance, and direction setting," Baybayan said. "It's based upon seeing the whole field of stars as it rises between the eastern and western horizon[s]." It's not the same as navigation, which is based on math and science. "They are different in the way they come to their knowledge base," said Baybayan. "Wayfinding is more about synthesizing the body through the pulse of the universe as she moves through the ocean. They both accomplish the same thing. But they take a different way of knowing to come to the same objective."

Baybayan took all these voyaging lessons back to land. "The whole process of navigation is about having a dream or a vision—an objective—and then creating a plan to get to the objective, and then defining all the skill sets you need to accomplish it, and then training for it, and then going out and executing it, and understanding that you need to be flexible along the way," he said. "Because the winds aren't always going to be blowing your way."

KĀLEPA BAYBAYAN—who grew up in Lahaina, Maui—was the head of the ʻImiloa Astronomy Center of Hawaii. He found his way there after his first sailing on the *Hōkūleʻa* in 1975 and continued to travel on the canoe in the decades that followed. He brought wayfinding to more people, and was inducted into the order of Pwo, which the Polynesian Voyaging Society describes as a "two-thousand-year-old society of deep-sea navigators in Micronesia."[4] He passed away in 2021.

V. THE PLANETS

IN WHICH YOU OBSERVE SOME OF THE MOTIONS OF OUR SOLAR SYSTEM COMPANION

OBSERVING A METEOR SHOWER*

CONJUNCTION JUNCTION IN THE SKY**

VENUS, THE PLANET THAT FLEW TOO CLOSE TO THE SUN**

HOW THE PLANETS MOVE RELATIVE TO THE STARS***

The solar system has eight planets. Telescopes and probes have shown their landscapes to be swirly or mountainous or cavernous or icy or dirty or red or brown or paisley-ish. They were formed as particles of dust—left over from the Sun's formation—crashed into each other, stuck together, grew via gravity, and eventually pulled themselves into spheres. Some of the planets have moons that rival other planets in size; other moons shoot geysers into space. Earth may be the only place with humans, but it has no monopoly on personality.

Earth and these seven other orbs (plus Pluto and a growing list of far-flung dwarf planets) orbit the Sun and move across our sky in ways different from stars. Five of them—Mercury, Venus, Mars, Jupiter, and Saturn—are visible to the naked eye. But unless you're out there night after night, noticing them on the regular, it's hard to see how their motions differ from those more distant objects. And because they look like brighter, sometimes more-colorful stars, they often get mistaken for faraway suns—or, more often than you'd think, for UFOs.

OBSERVING A METEOR SHOWER

You know the excitement of catching a surprise "shooting star" in the sky. Maybe you saw one on an evening stroll, or driving home after-hours. Maybe you made a wish, thinking of the old superstition that they're good luck. The idea behind that tradition is that these are rare events, and so the moment deserves a special thought. Plus, it's cool: it looks like one of the actual stars has in fact pulled up roots and zipped away.

As you probably know, these beautiful, intermittent streaks are not stars at all but in fact are often tiny pieces that comets have shed over time. Earth's orbit regularly crosses the debris fields of these comets. Their shards then rain down on the Earth, briefly and intensely heating up a path through the atmosphere, and lighting up a streak in the dark sky fifty to one hundred miles above your head. The vast majority of meteor streaks happen because of tiny "meteoroids," which range in size from a grain of sand to a shelled peanut.

If you happen to see a single meteor, the pattern might not be obvious. But each meteor shower has a point it radiates outward from. It's like driving in a snowstorm at night. As you white-knuckle along the highway, the snowflakes will all appear to radiate from a point ahead of you (please, make sure you're a passenger when you look for this effect).

At the end of this exercise is a list of meteor shower dates and what constellation they will appear to radiate from. Remember that showers are highly variable—you may see a lot, a few, or none, depending on how recently the comet was in the part of its orbit that Earth is about to plunge through. Worst-case scenario, you'll have some time lying on a blanket or sitting in a lounge chair, looking at the dark night sky. And no matter how many meteors you see, take a moment to think, as Kurt Vonnegut once said, "If this isn't nice, I don't know what is."[1]

WHAT YOU'LL NEED

A dark, clear nighttime sky
This book (or the internet) to determine your observing dates
A pencil or pen
Your Star Wheel (page 187)
A blanket or reclining outdoor chair
Mosquito repellent and/or warm clothing (depending on the time of year)
A notebook or sketchbook (optional)

WHAT YOU'LL DO

You'll pick a date from a list of popular meteor showers provided at the end of the chapter (page 157), and find a location where you can safely look at a large swath of the sky late at night. Find a place where you can be comfortable lying or sitting in a reclining chair. You'll observe the sky and make a note of how many meteors you see in a particular period of time. You can also draw the constellations that you can see and the approximate locations of the meteors.

THE DETAILS

Meteor showers occur throughout the year, so you should be able to do this exercise at almost any time (give or take a few weeks). Keep in mind that the best time to observe meteor showers is after midnight and when there is a New Moon, when the sky is darker. So stock up on coffee and check out a calendar.

- Using the table at the end of this exercise, or the internet, pick a night (or range of nights) when you want to observe. The American Meteor Society has a regularly updated web page that will display the best meteor shower dates in a particular year. Once you've chosen your night, confirm that a Full Moon won't stymie your efforts. Then note the name of the shower and the constellation it is associated with.

- If you're observing in the summertime and there are mosquitoes, make sure to slather yourself in some sort of repellent. If you're observing in the wintertime, put on a coat like your mother told you.

- Find a location where you have the most open view possible. An athletic field or open meadow is perfect. Meteor showers can be harder to observe in large cities, but a building rooftop or park may work.

- Using your Star Wheel, find the constellation from which the shower will appear to emanate. For example, if you're observing the Geminids in mid-December, find where Gemini will be floating at midnight.

- Note your observing date and time in the following space provided.

- Take a deep breath, look at the sky near your constellation, and make a simple drawing of the stars you see in that part of the sky.
- Watch, wait, and enjoy this moment looking out into the solar system and beyond.
- When you see a meteor, note its approximate position and time on your simple drawing.
- Watch for as long as you like. Try not to fall asleep.
- Repeat this activity throughout the year as often as the mood strikes, remembering always that your planet is passing through the debris of an ancient member of the solar system, whose constituent compounds helped make Earth what it is today!
- Make a wish. (Why not? Can't hurt.)

KEEPING IT SHORT

The investment of time for this exercise can make for a lovely night. But if you don't have a full evening to dedicate to observing, just pick a date during one of the meteor showers listed, and find a dark sky to watch for fifteen to twenty minutes. Lie down and get comfy, and look up at the sky. You might get lucky!

THINGS TO PONDER

- ⊙ How did you feel looking at the stars when you didn't see a meteor? Did that change when you saw one?
- ⊙ There are other objects you might see when you look at the night sky for this long. Did you see any? What were they?
- ⊙ In addition to meteors, you might see fireballs, or "bolides," which are generally the result of larger objects coming into the atmosphere. Have you ever noticed one of these events?
- ⊙ Was it easy or difficult for you to focus on the night sky? What did you find yourself thinking about as you watched for "shooting stars"?

MINDFULNESS EXERCISE (CAREFUL OBSERVATION)

Find a small, edible object, like a blueberry, a grape, or a peanut. This is the size of some of the objects that we see as shooting stars. Close your eyes. Notice its weight, how it pushes on your hands and how your hands push back on it. Is it warm or cool to the touch? Is it smooth or rough? Take a few minutes to focus on these sensations. Don't be concerned about the causes of these sensations, but focus on the sensations themselves: the weight, the temperature, the texture of the object. Now eat it mindfully, enjoying its unique textures and flavors as you slowly chew.

METEOR SHOWER DATES

(Check the American Meteor Society website for exact dates each year.)

METEOR SHOWER	DATES OF MAXIMUM ACTIVITY	NEAREST CONSTELLATION TO ITS LOCATION
Quadrantids	January 3–4	Along the ecliptic from Cancer to Virgo
Lyrids	April 22–23	Lyra
Eta Aquariids	May 7–8	Aquarius
Alpha Capricornids	July 27–28	Capricorn
Southern Delta Aquariids	July 29–30	Aquarius
Perseids	August 12–13	Perseus
Southern Taurids	October 11	Taurus
Orionids	October 21–22	Orion
Northern Taurids	November 13–14	Taurus
Leonids	November 17–19	Leo
Geminids	December 13–14	Gemini
Ursids	December 22–23	Ursa Major

CONJUNCTION JUNCTION IN THE SKY

Celestial objects, like pretty much everything else, *appear to* get intimate with one another on occasion, as they approach closely and comfortably in the sky. When two astronomical bodies *seem to* be near each other from our perspective on Earth, it's called a "conjunction." Conjunctions can happen between a planet and a star, a planet and a planet, the Moon and a planet, multiple planets and all those companions, or an asteroid or asteroids and those companions. In late 2020, Jupiter and Saturn got closer to each other at night than they had in eight hundred years.

The stars and other very distant objects, out beyond the solar system, move together in the sky, rising and setting in almost the same positions relative to one another. But the closer objects in the solar system don't hew to that same schedule; they move relative both to one another and to the farther-off stuff. That's why they can end up *looking like* they're creeping up on other sky things.

You may have noticed a few italicized words in the previous two paragraphs: *appear to*, *seem to*, *looking like*. What you see in the sky, where you see it, and when you see it depend completely on your perspective. Our view is a chronological and spatial coincidence. Constellations? Just a coincidence, based on the particular perch from where we look out on the stars in our galaxy. Conjunctions? Just a coincidence. Our experience of what's above us is—like almost everything else—subjective, and (cosmically) personal.

The objects that veer close to each other from our viewpoint aren't *actually* close to each other in space. It's a sort of visual illusion, like when an airplane flies in front of the Sun, or when you block the Sun shining in your face with your hand. Your hand isn't anywhere near the Sun. It's just in the same direction as the Sun (from your perspective). And that plane isn't Icarus.

That fact doesn't make conjunctions less cool to look at. In fact, it might make them cooler—because they're a special show for Earth, for you.

In this exercise, you'll check out a conjunction—a coming together—of two planets of your own choosing.

WHAT YOU'LL NEED

Access to an internet browser
This book, a notebook, or a sketchbook
A pencil or pen
A night-sky app (optional)
A clear nighttime sky

WHAT YOU'LL DO

You'll look up a calendar of when planetary conjunctions are set to occur in the near future, then go out and find one in the sky.

THE DETAILS

- Use your preferred internet browser to search for "planetary conjunctions + [insert year here]." Skip the ones that involve the planets Neptune or Uranus, which aren't bright enough for your eyes to see without a telescope. You should come up with several options of when different planets will appear near each other in the sky, including the time of evening. Pick your favorite.

- On the appointed night, at the appointed time, go outside and look up.

- Try, first, to see if you can find these twinning planets without using a night-sky app (they will be two bright things super close together—and might look unusual among the other stuff up there). If you can't, power up that app. Draw a picture of the planets in the space provided, including whatever else you notice on the horizon, for directional reference.

KEEPING IT SHORT

Look up a calendar of planetary conjunctions, and go out on the evening of your ultimate favorite. Using an app or just your eyes, find the two nearby objects and think about how even though they appear so glued to each other, they're still spread across a huge amount of space.

THINGS TO PONDER

- Conjunctions are rarer events, like shooting stars. But unlike shooting stars, they are predictable! Why is that? Do you get more of a zing from seeing something in the sky that you know is coming or something that's a surprise? Why?
- Why don't stars have conjunctions with each other?
- Do you think there are conjunctions between planets that you can't see because they happen during the daytime?
- What role do you think coincidence has played in your life, and how important do you feel it is in the universe?
- In your original sketch, also draw the orbits of the planets you're looking at, and Earth's orbit. Imagine yourself looking out from Earth. Where would the two planets have to be relative to each other to look close together in the sky? Make it as artistic, or not, as you want.

MINDFULNESS EXERCISE (CAREFUL OBSERVATION)

While you're watching the planetary conjunction, reserve a few moments to do a truly mindful observation of them. Don't concentrate on anything other than noticing these two points of light. As thoughts arise that aren't about these solar-system objects, let them go. Relax and simply watch the planets, trying to take them in as if you've never seen them before, although you've likely seen them many nights of your life (perhaps without realizing what they were). Observe whatever you can about these objects and their differences in appearance (their color, their brightness), as viewed from this specific spot on Earth at this specific moment.

VENUS, THE PLANET THAT FLEW TOO CLOSE TO THE SUN

Venus is a similar size, density, and composition to the Earth, but—at least in terms of how humans would fare there—it's a strange, hostile world. An evil twin, of sorts. The pressure there is about the same as you'd experience (dead) 3,000 feet beneath the ocean's surface. The average temperature is a body-melting 864 degrees Fahrenheit!

But even if Venus isn't a welcoming world, it's an interesting one. It spins the opposite direction from most other planets—and very slowly. Its "day" (one rotation) lasts 243 Earth days and takes longer than its "year" (one trip around the Sun), which lasts 225 days. The Sun rises twice during its year, traveling from west to east. In all that time, no moon traverses the sky, because Venus doesn't have one.

Humans have noticed Venus for millennia, given that it's the brightest planet in the sky, and the brightest object besides the Moon. But Galileo was the first to look at it through a telescope and see that it—like our Moon—has its own phases, going from Crescent to Full and back again.

You can easily see Venus just before sunrise or just after sunset, which led ancient astronomers to call it the morning star and the evening star. They did not necessarily realize that those two "stars" were one single thing—a thing that wasn't a star at all.

You, a modern person, know better. But while you've probably noticed that bright orb hanging near the horizon before any of the *real* stars come out, this exercise gives you the chance to think about why it appears there when it does, and think a little bit differently about its twinkle (which leads people to often mistake it for a UFO!).

WHAT YOU'LL NEED

This book, a notebook, or a sketchbook
A pencil or pen
A night-sky app

WHAT YOU'LL DO

You'll make a drawing of where Venus appears on the horizon and then think about the inner solar system and how the planets might be arranged so that Venus appears where it does, when it does.

THE DETAILS

- If you're up and outside near sunrise or sunset, check your night-sky app, or search online for Venus's rise and set times to find out where to look for it.

- If it's observable, look at it! What do you notice about Venus's color? Does the planet change in appearance at all as you watch it? Make a drawing of what you see, maybe something like the following image. In this drawing, the Moon is moving past Venus in the western, evening sky:

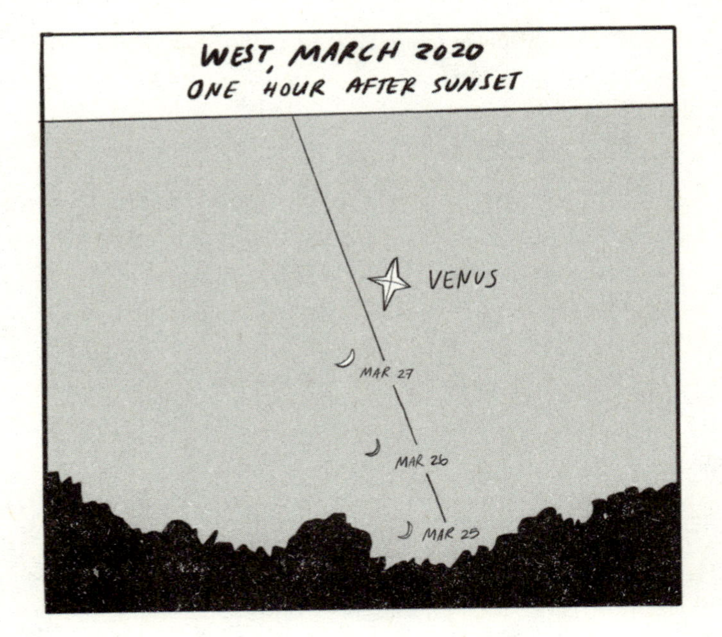

⊙ In the space provided, draw four concentric circles.

- Label the smallest circle at the center "Sun" and the three outer circles (starting with the next smallest one) "Mercury," "Venus," and "Earth."

- Try to figure out where Venus and Earth might be in their orbits, and draw an arrow pointing to where on their respective circles. Your drawing might look something like this:

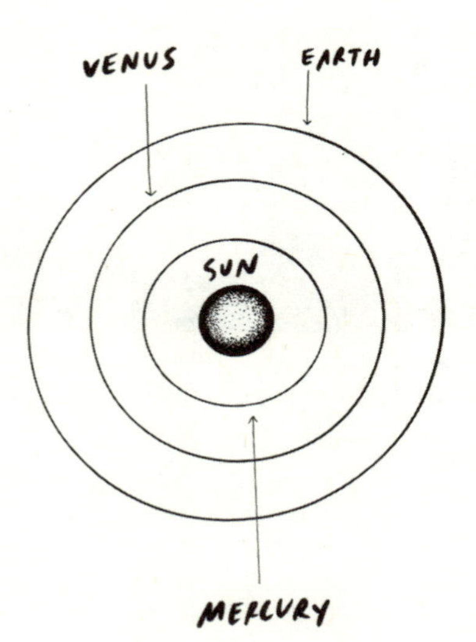

KEEPING IT SHORT

Using the internet or a night-sky app, see when and where you can find Venus (near sunrise or sunset, depending on whether you're a morning or evening star yourself). Imagine Venus circling the Sun in an orbit smaller than Earth's, always between us and the Sun.

THINGS TO PONDER

- What would it be like to live on a planet whose day is similar in length to its year?

- What would it be like to live on a planet with no moon?

- Would you feel differently about the solar system if there were no other planets between Earth and the Sun?

- The planet Venus was associated with warfare in ancient Maya culture. Many Maya buildings were arranged to highlight the relative positions of the Sun and Venus. If modern buildings were arranged according to celestial happenings, would you think of either the buildings or the happenings differently?

- Some of the earliest writings known (c. 1700 BCE, Babylonia), set down on clay tablets, record the rising time of the planet Venus and when it was first visible before sunrise and after sunset. Why do you think this point of light was so important to the writers?

- When you're looking out from the Earth toward Venus and the Sun, think about whether there's an arrangement in which they wouldn't appear near each other in the sky. Can you find one? It turns out, since Venus is on an orbit smaller than the Earth's, it can never get very far from the Sun!

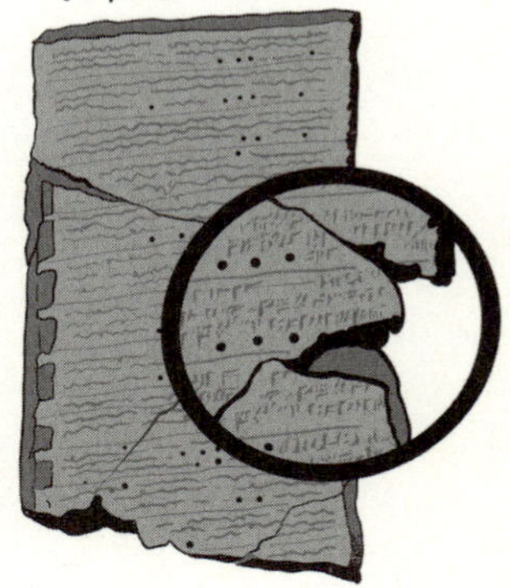

- Now think about the Sun rising and where Venus would look relative to it at different spots in its orbit. When would they appear close together in the sky? It turns out that Venus will always be just a little east or west of the Sun in the sky, meaning you will always see it just after sunset or just before sunrise!

MINDFULNESS EXERCISE (VISUALIZATION)

You don't actually want to live on Venus (it wouldn't go well), but lie down and imagine that you're somehow alive there, flat on its surface. Feel the intense pressure of its atmosphere pushing down on your skin, as if it's an airy compression suit, or an atmospheric weighted blanket. Imagine the intense heat on your body. Notice how it feels on your feet, then your legs, and all the way up to your head.

HOW THE PLANETS MOVE RELATIVE TO THE STARS

People noted long ago that some special stars were relatively bright and moved differently from the crowd. The Greeks called these stars *planetes*, or "wanderers." It was clear that somehow these *planetes* were different from the "fixed" stars in the night sky.

We now know that these objects are not stars at all but instead planets: other objects that go around the Sun in the same invisible pancake as Earth. The planets of the solar system orbit the Sun, and the Sun and all its planets make their way around the center of the Milky Way galaxy in a dance that is billions of years old.

Earth is the third planet from the solar system's center, and the farther a planet is from the Sun, the more slowly it orbits. So the closest planet to the Sun, Mercury, orbits every 88 days, Venus every 225 days, and Earth every 365 days. Planets farther out move even more slowly, and out at Jupiter's distance, it takes almost twelve years to trek around the Sun once. Saturn takes twenty-nine years to orbit. If you observe the night sky over a few carefully chosen nights, you can watch the journey of some of the planets closest to the Sun.

In the solar system, with Mercury and Venus closer to the Sun than we are (see the following picture), they are never very far from the Sun in the sky.

Imagine watching someone swinging a yo-yo around them above their head. You're watching the person swinging the yo-yo, and you're far enough away that you don't get hit! The person is the Sun, the

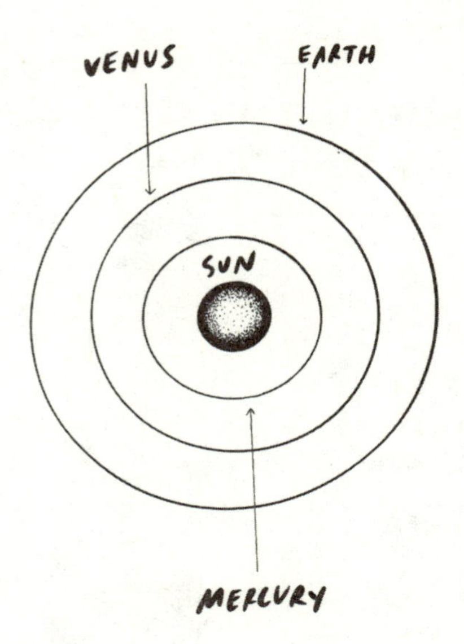

yo-yo is Venus (or Mercury), and you are Earth. As you watch that yo-yo swinging around, you will notice that it never gets any farther away from the person than the length of the string.

The same effect is seen with the two planets closest to the Sun. They are on shorter "strings" than we are, so as we watch them, they never venture very far from the Sun. Mercury, following the shortest path, stays within about three clenched fists—if you hold your arm all the way out and your hand against the sky—from the Sun. Venus never gets more than about five clenched fists.

When Venus is most distant from the Sun (on one side of its yo-yo swing), it appears about halfway between the western horizon and overhead at sunset. The next drawing shows where Venus and the Moon were relative to each other from March 25 to March 27, 2020. Notice that over the course of just a few days the Moon moved pretty fast (about the width of one outstretched hand, held at arm's length, per night), while Venus doesn't appear to have moved much at all. You'll have to watch a planet like Venus for more than a few days to see it "wandering" among the stars.

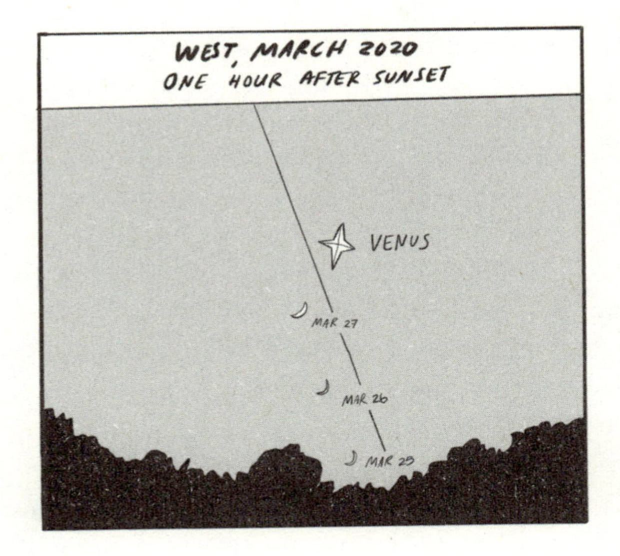

WHAT YOU'LL NEED

A clear nighttime sky
This book, a notebook, or a sketchbook
A pencil or pen
The setting time of Venus in your location
A few weeks to a few months to make observations

WHAT YOU'LL DO

In this exercise, you'll carefully observe the position of Venus as it moves across the nighttime sky (post-sunset, pre-dawn, or both). To do this, you'll need to know the setting and/or rising time of Venus from your observing location. At that point, you can look at the planet every few days and follow its subtle motions through the sky.

THE DETAILS

At the outset of this exercise, it's good to remember that Venus's travels will be a little harder to observe from night to night than the Moon's. Venus moves only about 1 degree per day against the background stars. The Moon, on the other hand, moves about fifteen times farther! Our goal is to catch Venus when it is farthest from the Sun (keep thinking of the yo-yo, swinging over the head of our strange friend) and follow its motion in the sky. It will be easier to follow Venus if it is near some familiar constellations, so you can notice its motions more easily.

Note also that when you check for Venus's rising and setting times, if it's not slated to set two to three hours after the Sun sets, or rise two to three hours before the Sun, it's best to wait to start this exercise. The longer the time between sunset and Venus setting, or sunrise and Venus rising, the more time you'll have to observe the planet in darkness.

- ⊙ Pick a time about thirty minutes after sunset to make your observation.

- ⊙ Sketch your southwestern horizon in the space provided, noting any local landmarks. On the first day of your observation, make note of any constellations or patterns of stars in the southwestern sky.

- Place Venus (which should look very bright) in your sketch of the sky relative to landmarks and any patterns of stars you can see.

- If possible, go out for the next few days and draw the new position of Venus relative to those nearby stars and your local horizon.

- Repeat these observations as often as you can (maybe twice a week) for the next month. Over time, you should notice that the position of Venus, relative to the background stars and the horizon, is shifting.

- Make these observations for as long as you can. Eventually, Venus will get so low in the sky that you will not be able to observe it because of trees, houses, and tall people on your horizon.

KEEPING IT SHORT

You can still see Venus in motion if you watch it less often and over a shorter time. It will still be essential to catch Venus when its setting time is two to three hours later than the Sun's. Note its position relative to the southwestern horizon as often as you can for about a month. You should be able to see Venus getting closer and closer to the Sun!

THINGS TO PONDER

- That yo-yo swings around both sides. If you're an early morning person, then you could do this same thing before sunrise, as mentioned earlier. You'll just need to check that Venus is rising about two to three hours before the Sun. That's right: two to three hours before sunrise.

- Have you ever noticed a bright object on the eastern horizon as you were driving (east) to work in the morning? It might have been Venus!

- Do you have a planetarium near you? If you do, go visit it, and see if you can get a demonstration of these motions!

- Venus might just look like a bright point in the sky, but what changes for you when you really focus on the idea that it's a planet as large as Earth that may once (billions of years ago) have had oceans and even life?

- In a lot of twentieth-century science fiction, Venus was portrayed as a tropical planet, swept by rains and covered in vegetation. In fact, it is a very hot, inhospitable place. Why do you think old sci-fi writers imagined it in this way?

MINDFULNESS EXERCISE (VISUALIZATION/BREATHING)

Close your eyes and imagine that you have a bird's-eye view of a planet slowly orbiting a star in a circle, again and again. As the planet orbits the star, breathe in slowly as it goes halfway around the orbit. Then breathe out slowly as it travels through the other half. Slow and deepen your breaths, imagining the planet moving more slowly in its orbit around the Sun as you lengthen your breaths. Continue this exercise for ten deep breaths.

INTERLUDE:
THE ASTRONOMICAL ALIGNMENTS OF CHACO CANYON

As you walk through [Chaco Canyon], you are in the midst of . . . the pragmatic, the scientific, and the mystical.

—DAVID WARREN, ETHNOHISTORIAN, SANTA CLARA PUEBLO[2]

The archaeological record reveals what you probably know by now: people have pretty much *always* been mindful of what's going on above their heads.[3] You can find stone structures the world over—both modest and grand—that align with particular points on the horizon. The prevalence of these structures, and the herculean effort that went into moving stones and timbers hundreds of miles, indicate that their makers cared deeply about noticing the sky and marking certain dates.

Despite the chronological gap between these builders around the globe, their skies had similarities. No matter where you find yourself in time or space, the Sun's setting site always tells you what time of year it is. The Moon's phases always repeat every 29.5 days. That's close to, but not quite, a modern month, so the Moon's phases drift through our twenty-first-century divisions in the calendar. Some cultural and religious traditions, like Islam and Judaism, base their calendar on the Moon and start a new month with each new lunar cycle.

Indigenous peoples in Central and North America also marked time this way. Located in current-day New Mexico at over 6,000 feet in elevation, Chaco Canyon was a cultural hub for the ancestors of today's Pueblo Indians. Humans had lived there since at least 3,000 BCE. But at around 850 CE, they built massive, dramatic structures—now in ruins. Many of the buildings seem to line up with the compass points and with one another along astronomical axes, even though they're separated by great distances. The 1999 film *The Mystery of Chaco Canyon* explores these ideas more fully.

While hunting down ancient rock art on Fajada Butte in the 1970s, archeoastronomer Anna Sofaer discovered something remarkable: three slabs of sandstone leaning against a cliff wall. She happened to be visiting close to the summer solstice. On that day, an opening in the stacked slabs allowed a shard of light to illuminate the dark space behind it. A bright band spread from top to bottom across a spiral, carved painstakingly into the rock behind the slabs. At noon local time, a dagger of light stabbed perfectly through the middle of the spiral.

As Sofaer kept watch throughout the year, she saw that on the equinoxes, in March and September, one dagger of light illuminated the large spiral and another lit up a smaller pattern to its left. On the winter solstice, a pair of daggers bracketed the large spiral, like bookends.

Soon Sofaer started the Solstice Project organization and made careful observations of the structures in Chaco Canyon, with the help of the National Geodetic Survey, a project of the National Oceanic and Atmospheric Administration (NOAA).[4] It turns out, the light dagger was just one in a collection of careful lineups. The largest building has long, straight walls that line straight up with north–south and east–west, like roads in some modern cities. The east–west walls are shadowless at sunrise and sunset on the equinoxes, and the north–south wall casts no shade at noon.

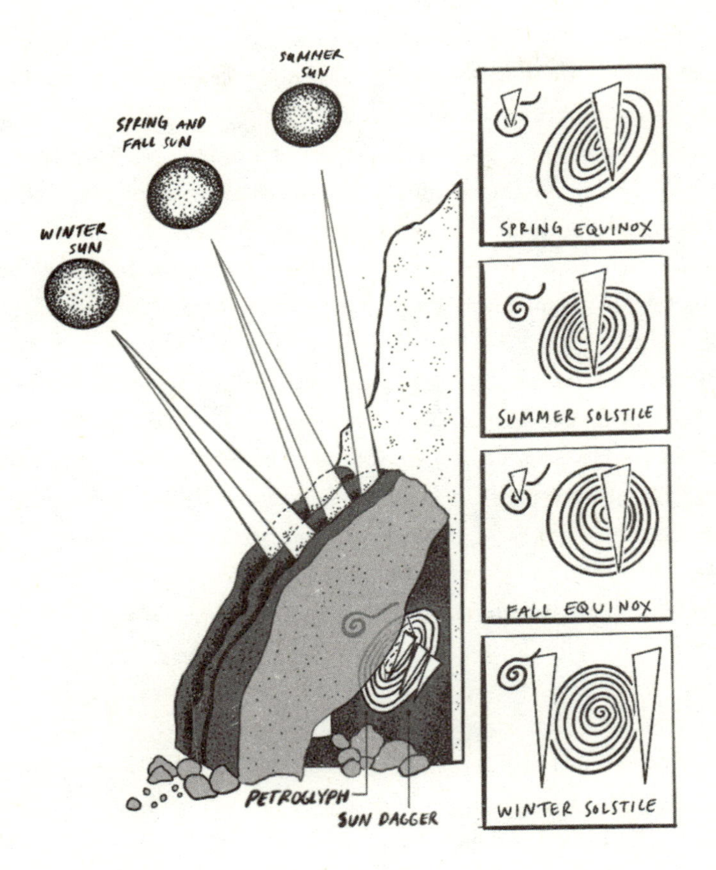

Daggers of light illuminate different parts of the carved spirals on different important days each year.

Phillip Tuwaletstiwa—Hopi Indian and former deputy director of the National Geodetic Survey—joined the Solstice Project to learn more about the buildings and their relationships. His measurements showed that the ancestors of the Pueblo Indians were "close observers of nature."[5]

He continued: "If there was a way to transfer the orderly nature of the cosmos down onto what seems to be the chaos that exists here . . . then you begin to integrate both Heaven and Earth."[6]

In the decades since the Sun Dagger's discovery, the Solstice Project confirmed that some Chaco Canyon structures are also engineered to reflect the Moon's location. At the Sun Dagger, Sofaer and her team found that on the night when the Moon rises farthest to the south, its Full shadow slices the larger spiral in half. Exactly 9.25 years later, the shadow darkens the edge of the spiral. On top of that, if you stand at the central building, called Pueblo Bonito, surrounding structures match these lunar locations.

The ruins of Chaco Canyon were declared a UNESCO World Heritage Site in 1987, and in 2013, Chaco Canyon, with its pristine celestial views, was declared an International Dark Sky Park. If you're visiting New Mexico, it's well worth a stop. But bring a hat, sunscreen, and a lot of water.

INTERLUDE:

PROFILE OF CULTURAL-ASTRONOMY EXPERTS WILLY BARRENO AND ISABEL HAWKINS

"I was born in the middle of war in Guatemala, and my people were very oppressed," says Willy Barreno, cofounder of the Ki'kotemal School for young Maya.[7] "We didn't know we were Maya when we were growing up, and we were ashamed of our culture. I remember, growing up, just dreaming that I was an Egyptian because I wanted to be someone else. I was always admiring the pyramids and the stars and the mythology. I thought it would be so cool to be an Egyptian because they were smart."

Today, he wants Maya kids to instead learn about and admire their *own* heritage—astronomical and otherwise.

Barreno began to investigate his past more fully only when he came to the US as a refugee and lived in New Mexico for seven years. "It was very important to be in New Mexico because the Pueblo people were really curious about who the Maya were, and I was pushed into learning my own culture." Getting to the astronomy part took longer, though. "I used to go camp at Mesa Verde in Colorado, and then I started seeing the stars, but I still didn't know much about my own culture in terms of astronomy. I knew we were setting up fires and counting the days, but I didn't know much about the knowledge my ancestors had created thousands of years before."

He's gone on to find much more of that insight, and to pass it onward. Students at Ki'kotemal School have locked on to how their ancestors' view of the sky differs—literally and figuratively—from that of other cultures. "I always thought there was a Milky Way up there," says Barreno, "but remember that most of astronomy and astrology is based on Greek mythology. *My* ancestors called it the white alligator. They were looking for an alligator because that's what we had in the jungle, not a Milky Way."

Besides, he adds, the Maya are lactose intolerant.

And in Orion, they see not a belt but a triangle. "That's from the sacred fire from the creation of the Maya," says Barreno, who started studying Maya spiritual practices more when he moved back to Guatemala.

In 2010, at one spiritual ceremony, he met an astronomer named Isabel Hawkins, who currently works at the Exploratorium science museum in San Francisco, California, but grew up in Argentina. Hawkins and Barreno now collaborate and share knowledge in "cultural astronomy," which she calls the "astronomy of well-being."

"Two important moments in my childhood led me to astronomy," Hawkins says. "One was growing up on a ranch, and every summer we'd spend there. It's a place that's still very isolated in northern Córdoba. It's very hot and dry in the summer. My cousins and I would pull out the mattresses and sleep outside. The sky was just blanketing you very beautifully, especially the Southern Hemisphere. There's all these shooting stars constantly crossing the sky."

In the morning, she and her cousins would go looking for the shooting stars, searching for where they had fallen. "Which—thinking back—is an astronomical observation," she says, "the relationship between the sky and your particular place."

When Hawkins was ten years old, she visited a planetarium and was amazed at the director's ability to move the sky and make sense of celestial patterns. "From then on I wanted to understand why the universe was the way it was, and why the sky was so dark between the stars," she says.

"I ended up studying the interstellar medium, the stuff between the stars."

She became involved in cultural astronomy, though, through NASA—and the closest star. "They contracted me at UC Berkeley, when I was at the Space Sciences Lab, to do work for the public, to share heliophysics, or the science of the Sun," Hawkins says. "We started looking into total solar eclipses as a means of sharing the dynamic nature of the Sun, and we also started to work with colleagues in Mesoamerica—particularly Mexico and Guatemala—to highlight the legacy of astronomy from our ancestors in the Americas."

That's how she met Barreno, and then learned of his work with the Ki'kotemal School, which helps portray the most fundamental idea of Maya astronomy. "Whatever happens in the sky happens in the Earth," says Barreno. Students learn that, in part, through an agricultural program that teaches astronomical tradition as it relates to planting and harvesting.

"The teaching about the stars is always in the context of the teachings about the Earth," says Hawkins. "It's all connected. When the children come to participate in the ceremonies for the planting and harvesting of the corn, there's this conversation that integrates the movement of the Sun, the Moon, the planets, and the corn and the other important plants on Earth. It's an astronomy of connecting and relating the cycles of the sky in service of the agriculture, and then the agriculture responds to what goes on with the Sun, the Moon, and the planets. There's this reciprocity that's so important."

Beyond that interplay, says Barreno, astronomy teaches us to look backward. "Seeing through the depths of the universe, you are seeing also the past," he says. He wants modern Maya to do the same: to understand their present, personal universe by looking back in time to their ancestors.

His work uncovering the past is key to passing knowledge on. "Since the colonizers came here, they burned all the books," he says. But he thinks that within a decade, they'll have recovered all the information they need to keep track of the sky the way people there did thousands of years ago.

"Then the next generations of kids will—just with their fingers and just a little bit of basics—keep track of eclipses and solstices and equinoxes again," he says.

Barreno doesn't, though, think those kids need to stick only to old methods. "We also need to adjust to modern technology if we want to see the future." A good balance, he says, is exemplified by a single mom to whom he's teaching Maya day-keeping practices. Barreno gave her son a night-sky app for his phone. "The kid is in love with that, and every day he doesn't go to sleep," says Barreno. "He sees the stars in there."

When the child's mother took him to a fire ceremony, he asked for a telescope. The modern and its past, in other words, can coexist peacefully. And having a science rooted to your location, and your own culture, does more than reveal the location of a planet on a given day. "The astronomy of place, the astronomy of relationships, fosters your identity," says Barreno.

Photo courtesy of Willy Barreno

WILLY BARRENO is originally from Quetzaltenango, Guatemala. At age twenty-three, he immigrated to the United States, where he worked as a chef and community organizer in the area of human and Indigenous rights in different states. After twelve years of living in the US, he decided to return to Guatemala to recover his identity and rediscover his roots. As part of his current work, Willy is a cofounder of the binational network DESGUA, which promotes fair trade and social enterprise initiatives in Guatemala. He is an *aj'q'ij* (Maya calendar keeper) and helps run the Ki'kotemal School.

Photo courtesy of Isabel Hawkins

DR. ISABEL HAWKINS is an astronomer and a project director at the Exploratorium science museum. She obtained her doctorate in astronomy at UCLA in 1986 and conducted postdoctoral work at UC Berkeley as a member of NASA's Extreme Ultraviolet Explorer satellite mission. Isabel received a Fulbright Global Scholar Award in 2019 to research the Pleiades star cluster and intercultural connections in New Zealand, Guatemala, and Peru.

APPENDIX A:
ASSEMBLING AND USING YOUR STAR WHEEL

The night sky can seem overwhelming. There are just so many stars, and aside from a few well-known dot-to-dots, like Orion, the Big Dipper, or the Pleiades, you may have a hard time finding anything. With some practice, you'll get more and more familiar with the stellar groupings that speckle the sky. But you probably didn't learn to ride a bike or rob a bank overnight. It took time, and effort, and sometimes some training wheels.

There are easy ways to gradually become acquainted with the night sky and to know what's up there even when the view is marred by light pollution. Big-city lights wash out all but the brightest stars, and if you look up from Paris or Chicago or Hong Kong, you might see very few stars indeed. If you have a smartphone and want its assistance, check out appendix B (page 193), about that very topic. But what if your phone battery dies? Or there's a zombie apocalypse? Or the Global Positioning System (GPS) collapses for suspicious but unknown reasons?

That's where your Star Wheel comes in. This book contains your very own pocket guide to the night sky. All you need to assemble it is a pair of scissors, a steady hand, some tape, and about five minutes. Once put together, the Star Wheel will reveal what constellations are visible right now and how the sky morphs over a night, a month, or a year.

ASSEMBLING YOUR STAR WHEEL

You can find a variety of Star Wheels online to print out or purchase. This book comes with two Star Wheels of its own, but if you happen to buy this book used, and the Star Wheel is conspicuously missing, you can always download another Star Wheel from the Discovery Corner Toy and Book Store of the Lawrence Hall of Science or *Sky & Telescope* magazine.

The Star Wheel in this book is printed on thick paper, suitable for cutting and assembling. If you get a PDF online, make sure to print it on thick paper (not regular paper) to make it more durable and usable.

Follow the simple cutting, folding, and taping instructions printed on the Star Wheel. Basically, it has two pieces when completed: an envelope with an oval "window" that shows you the night sky, and the circular wheel itself. Assemble the envelope, and slide the circular wheel that you cut out into the pocket. When properly assembled, it should look like the image on the next page.

What the Star Wheel shows you is your local view of the night sky on a particular date and at a particular time. The Star Wheel included in this book is for the Northern Hemisphere and is relatively accurate from 30 to 50 degrees north latitude. If you live in the Southern Hemisphere, you can load a different PDF and print it.

A Star Wheel shows you your local sky from horizon to horizon, and the center of the oval cutout represents the spot directly overhead—your zenith.

Your Night Sky Tonight

Let's go ahead and set your Star Wheel for a particular date. Rotate the wheel in the pocket until the date, let's say, April 20 aligns with the time you want to observe, let's say, 9 p.m. Inside the oval, the con-

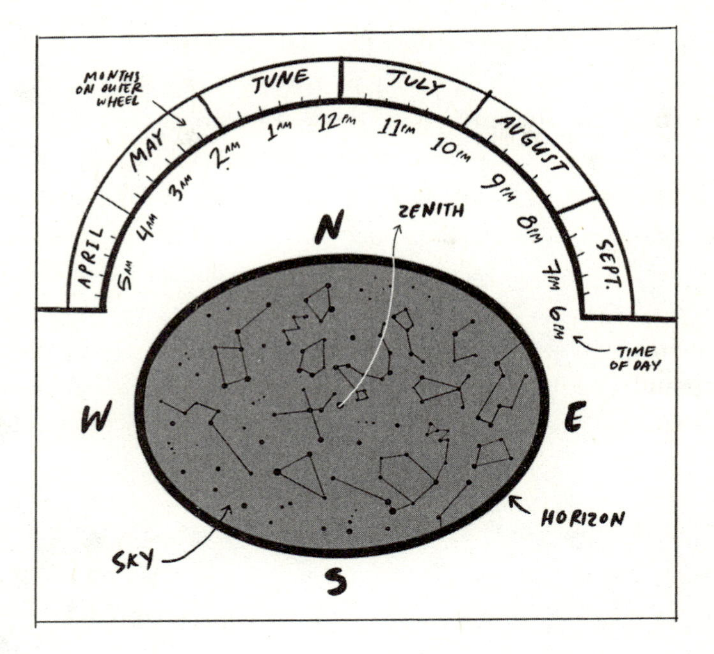

stellations splay across the sky. Groups of stars close to the western horizon (marked on the oval cutout) are about to set, and those close to the eastern horizon have just risen. Hercules has just risen in the northeast, and Orion is getting close to setting in the west. It looks like Ursa Major (the Big Bear, the bright stars of which most people call the Big Dipper) is almost directly overhead, near the center of the oval. Sometimes it helps to hold the Star Wheel overhead with the directions on the oval matching up with the cardinal directions on the horizon. That will help you to know where to look.

How the Sky Changes with Time

If you rotate the wheel counterclockwise, you're moving the hours forward. Time travel! In the previous example, you can look at the night sky on April 20 at 9 p.m., 10 p.m., 11 p.m., 12 a.m., and so on.

Stars and constellations continue to rise in the east and set in the west. By 11 p.m., the stars in Orion have mostly set, and by midnight, the constellation has disappeared for the night. New constellations, like Aquila the Eagle, are rising in the east. Keep rotating counterclockwise until you reach 5 a.m. on April 21. Hercules, which was just rising in the east at 9 p.m., now appears in the southwestern sky. Many of the constellations visible at 9 p.m. have set below the western horizon as the Earth has rotated.

Now let's look at this a different way. Any view of the night sky corresponds to a bunch of different dates and times, but there's a rhyme to this: The sky that you see at 5 a.m. on April 20 is the same

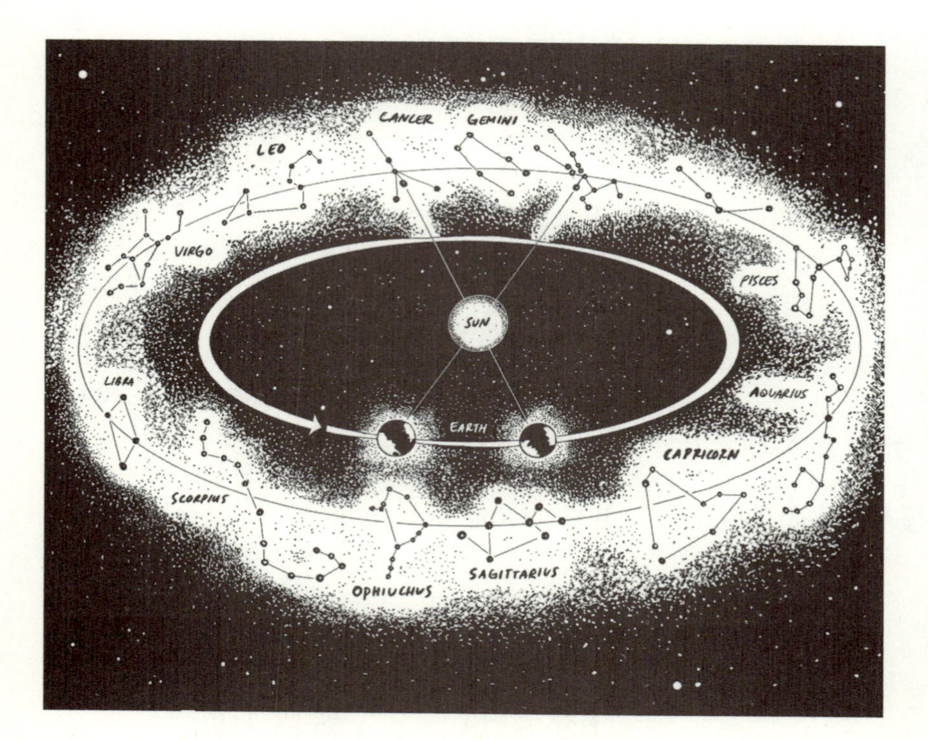

sky that you'll see at about 3 a.m. on May 20 or at about midnight on July 7. As the Earth orbits the Sun, the same sky shows up earlier and earlier in the evening throughout the year. And the stars that you can see best, burning through the middle of the night, are on the far side of the Earth, away from the Sun. As Earth orbits the Sun, those stars change. (This concept is explained in more detail in "Why Can't I See Orion Tonight?" page 135.) The image on the previous page shows this effect.

Armed with this knowledge and your Star Wheel, you're ready to take on the night sky. You'll probably need it less and less as you become a more experienced sky-gazer. You might eventually look at it fondly, like those training wheels that sit on a shelf in the garage, waiting for the next person who wants to learn to ride a bike. Now head out into the night and see if you can identify some star shapes!

APPENDIX B:
RECOMMENDED NIGHT-SKY APPS

Sure, there's something to be said for knowing your way around the sky without any outside help. That might give you a feeling of mastery over the universe, and perhaps over your life. But it's also a little like your dad saying he doesn't ever need to use GPS because he's lived here forever and knows his way around. That may be, but he could probably learn a trick or two from ceding some control to a smart, digital device.

The same is true of navigating the night sky. Apps can do quick celestial calculations your brain can't, can pinpoint your view exactly in a way your eyes—left to their own devices—would have to fiddle with, plus they can just make many of the exercises in these pages more fun by saving you some time and tedium. And if you decide to use any of these apps, you'll probably find over time that you need them less and less.

Listed here are a few apps you can use on either Apple or Android devices, with notes about their distinguishing features. Each of them will, once you point your phone at the sky, show you what you're seeing—constellations, stars, planets, the works. Download one, two, or grab them all! Use the one you like best.

Beware that the apps recommending GPS will only work as intended if your device has GPS enabled! At the time of this writing, all these apps cost less than five dollars.

STAR ROVER

GPS RECOMMENDED: YES

Star Rover includes not just all the constellations, planets, Moon phases, and deep-sky objects but also bonuses, like simulations of eclipses and details on any object in its vast catalog.

SKYSAFARI

GPS RECOMMENDED: NO

SkySafari can show you the normal natural objects in the sky (including asteroids and comets!) but also the satellites humans have launched to orbit around Earth. Plus you can ask it to show you animation of things like meteor showers and conjunctions.

SKYVIEW LITE

GPS RECOMMENDED: NO

This app has an augmented reality option that doesn't just display an illustrated version of the sky; it uses your phone's camera to capture a live image of the sky and overlays information on it!

COSMIC WATCH

GPS RECOMMENDED: YES

Cosmic Watch boasts a 3D view of the sky that shows the whole celestial sphere (or an "outside" view of the night sky) and keeps an ever-updated calendar of interesting sky events.

STELLARIUM MOBILE SKY MAP

GPS RECOMMENDED: NO

Stellarium is a planetarium in your pocket. It has 600,000 stars in its arsenal, compared to the more usual 120,000 or so. You can dial up any time, date, or location and see how the sky would look there and then.

STAR WALK 2

GPS RECOMMENDED: YES

You don't just have to see the sky as it is: you can get in a time machine and tell Star Walk 2 to show you what'll be above you on your birthday five years from now. Give yourself a present by using the augmented-reality capabilities to superimpose object names and outlines on the sky your camera sees.

STAR TRACKER

GPS RECOMMENDED: NO

If you want to keep your display simple but also have access to augmented reality and "night mode," which keeps your eyes adjusted to the darkness, try Star Tracker.

STAR CHART

GPS RECOMMENDED: YES

Want to see what's going on above people directly across the world from you—if you could tunnel through Earth and come out the other side? Star Chart's your go-to.

MOON PHASE CALENDAR

GPS RECOMMENDED: NO

Use this detailed calendar to see what's up with Earth's nearest neighbor, and how its phases and rise and set times will progress as the days go on.

APPENDIX C: HOW TO FIND ASTRONOMICAL IMAGES ONLINE

Looking at the sky with just your 20/20 or glasses-aided vision can be an awesome experience. Planets pierce the darkness, bright and twinkling. Multicolored stars arrange themselves into storied shapes. Under dark skies, the dusty Milky Way fuzzes a diagonal swath. Occasionally, the Moon shows its cratered face, lighting up the night.

But, really, our eyes aren't great astronomical observers. They can get the big picture, but they can't get detailed images, and they can't go deep. They also can only see light that comes from a very limited number of wavelengths. That, of course, is why astronomers have, for centuries, constructed telescopes, which detect everything from radio waves to gamma rays coming from space. With their lenses and mirrors and dishes, they can see farther and better, bringing distant objects first to scientists' own eyeballs, then to glass plates, then to detectors and cameras, and, finally, onto the internet.

Many large observatories—whether they're in space or on the ground—share their images openly. What follows are some ways you can find the kinds of celestial images your eyes will never be able to see, all for free:

A low-key way to see deep space is NASA's **Astronomy Picture of the Day** (apod.nasa.gov/apod/astropix.html). You don't have to know what you want to see: the space agency will just show you something different every day of the week. Visit the archive to dive into the past. Think of it as your curated daily dose of astronomy images. Many images also come from amateur astronomers.

If you want to choose your own adventure, you can try **NASA's image galleries** (www.nasa.gov/multimedia/imagegallery/index.html), which include a searchable library as well as mission-specific catalogs—so you can zoom in on the Hubble or Mars-rover shots that you love the best.

The European Space Agency (ESA) has a similar page for its pretty pictures (www.esa.int/ESA_Multimedia/Images), where you can search for whatever it is you seek.

If you're looking for snaps from telescopes rooted to the ground, consult the images at **NOIRLab** (noirlab.edu/public/image).

The **European Southern Observatory** (www.eso.org/gallery) has its own fine collection of free images, sorted by topics like "solar system" and "quasars."

On the other hand, if you're hankering for some more unusual (nonoptical) views, consult the **National Radio Astronomy Observatory**'s page (public.nrao.edu/gallery), where you can see what the universe would look like if your alien eyes saw into the radio part of the spectrum.

Travel south to find the **Australian Telescope National Facility**'s many radio telescopes (www.atnf.csiro.au/outreach/images/astronomical), whose images show planets looking like you've never seen them before, gas you can't see at all, and distant galaxies blazing with radiation.

The **South African Large Telescope** (www.salt.ac.za/galleries/photo-gallery) is also in the Southern Hemisphere, but it detects optical light, like a traditional telescope. Check out its greatest hits.

Run by NASA's Goddard Spaceflight Center, **SkyView** (skyview.gsfc.nasa.gov/current/cgi/titlepage.pl) calls itself "the internet's virtual telescope." You can put in either the name of an object or any coordinates you're interested in seeing, and select which astronomical survey's portrait you'd like to view. If you click on the "SkyView Query Form" in the left-hand menu, you can choose from even more surveys and change the size of the area you're seeing.

If you know the name of another observatory—perhaps one close to you—type it into your preferred search engine, add "image gallery," and chances are, something beautiful will pop up!

APPENDIX D: STAR BRIGHTNESS AND COLOR

Some stars shine much brighter than others. They also have a rainbow of colors. You can appreciate those personality traits in their own rights, but they also reveal a lot about what each star is actually like.

Lest we assume that people only started paying attention to color in the twentieth century, it's important to note that Aboriginal Australians have been noticing these properties for millennia. In a recent study, astronomer Duane Hamacher described two tribes in South Australia who have watched the varying brightness of several red-colored stars, including Betelgeuse (Alpha Orionis), for tens of thousands of years. Their oral traditions often reflect observations like these, with stories connected to the observed and repeated changes in the brightness of a star.[1]

Earlier in this book, the constellation Orion was mentioned, and the star Betelgeuse, found in the eastern shoulder of the hunter, is one of the brightest stars in the night sky. It makes the top-ten list, in fact! And Sirius A, the Dog Star, is the number-one brightest star, located close to Betelgeuse in the sky, in the adjacent constellation Canis Major, or the Big Dog. If you can see these two stars, try to distinguish their colors. Betelgeuse should look red, and Sirius A should be whitish-blue and brighter than Betelgeuse.

Since Sirius A catches the eye, you might assume that it's closer, and in this case you'd be right: it's only about 8 light-years away. And if all the stars had the same intrinsic brightness—if each were a 100-

watt light bulb—then their intensity would depend only on how far away they were, with the more distant "light bulbs" being fainter.

But stars come in a wide variety of "wattages," just like actual light bulbs. Sirius A is 25 times brighter than the Sun, but Betelgeuse is (wait for it) 126,000,000 times brighter.

So why is Sirius A so much brighter to our eyes?

Simply because it's closer. A lot closer. We may be 8 light-years from Sirius but we're 640 light-years from Betelgeuse.

Let's think of this another way. The Sun is the brightest star in the sky (so bright it gives you a burn), but it's actually just an ordinary star that happens to be basically on top of us.

And not all stars shine the same all the time. Many of them change, if you watch them closely. The most dramatic, sudden, and unpredictable changes happen when truly huge stars die in supernova explosions. In an instant, a star like Betelgeuse, already bangingly bright, can suddenly become almost ten thousand times more luminous.

Stars' colors are subtle but noticeable. On a clear night, you should be able to see, even with the naked eye, that they have a variety of hues. Noticing the palette can enrich your observations and help you to pay closer attention to the sky. As you gaze at the stars, your eyes are catching photons in a rainbow of colors that have been hurtling to Earth for tens, hundreds, or even thousands of years. Remember, stars that look red are cooler (relative to other stars, at least), and stars that look blue or blue-white are hotter.

Think about the heating element on an electric stove. When you first flip it on, it's cold and just looks grey. As it heats up, the element first gives off photons of light in the infrared part of the spectrum. You can feel its warmth but not see any light. Eventually, it heats up enough so that it gives off optical photons and glows "red-hot." If it gets very hot, the element will start to turn yellowish white.

Stars also give off light according to their temperature. Cooler stars give off the most radiation as infrared light and look reddish, while other stars give off the most light in orange, yellow, green, blue, or violet as they get hotter. The very hottest stars actually peak in the ultraviolet part of the spectrum and appear mostly white. Stars are clearly able to throw a lot of radiation toward their planets and the people (or aliens) on them.

So you can enjoy the variety in stars for its own sake. And with tools no more complex than your eyes, you can understand a lot about the inner workings of a sun located tens or hundreds of light-years away.

APPENDIX E: STAR NAMES

The names of stars often aren't the sorts of words people in English-speaking countries normally throw around. They sound unfamiliar, archaic, and downright strange. There's a good reason for that: their monikers were made up hundreds or even thousands of years ago. And while the bright stars have catchy names, like Aldebaran or Castor, which sound like things a celebrity might name a baby, many fainter stars have longer ones, like Alpha Centauri, or antiseptic-sounding ones, like HD 209458.

Cataloging the stars is an old business. As mentioned earlier, Indigenous Australians were among the first to note the positions and motions of stars over time. Their tradition was an oral one, but most of the named stars visible to the naked eye in modern-day astronomy also have far older Australian names. Ancient Egyptians, too, identified stars, but they made no formal catalogs that we know about.

The first known written catalogs of any sort come from ancient Sumerians, who wrote the names of a limited number of constellations on clay tablets. Babylonians recorded the first star catalogs about four thousand years ago, listing thirty-six in their who's who of the celestial equator.

It seems that many societies first named stars near the Sun's and the Moon's paths, which seemed somewhat special. In China, for example, astronomers appear to have grouped stars into "mansions" associated with the Sun as early as 200 BCE, while individual stars had names hundreds of years before that.

Ancient Greek astronomers like Eudoxus (c. 400 BCE) and Hipparchus (c. 100 BCE) built the first comprehensive star catalogs. And Ptolemy built on Hipparchus's work to make a list he called *Almagest.* One of the most famous Persian astronomers, Abd al-Rahman

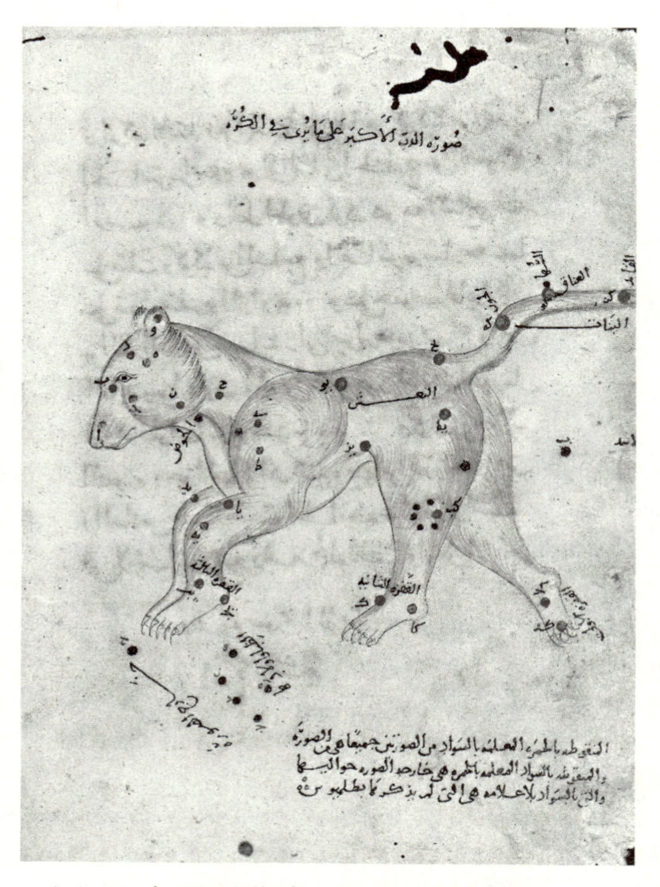

An image (c. 1009 CE) of the Big Bear (Ursa Major) from The Book of Fixed Stars. THE BODLEIAN LIBRARY

al-Sufi, wrote his improvements on Ptolemy's *Almagest* in Arabic and called the finished work *The Book of Fixed Stars* (c. 1000 CE).

All this history reveals why stars have names like Betelgeuse. Betelgeuse appears to be a westernized version of the Arabic description of the star's position (either hand or armpit) and its constellation (Orion). It's like a game of stellar telephone. Because astronomers

An illustration of the Orion constellation from Uranometria.
UNITED STATES NAVAL OBSERVATORY LIBRARY

writing in Arabic were updating and modernizing Ptolemy's ancient catalog, many bright stars took on Arabic names, and then westernized versions of Arabic names.

Over time, telescopes revealed even *more* stars, so astronomers needed to up their name game. That's why the catalogs of Johann Bayer and John Flamsteed—from the sixteenth and seventeenth centuries, respectively—contained particular naming schemes. Bayer's catalog, *Uranometria*, named the bright stars with a Greek letter, followed by the name of the constellation they were in. For example, the bright star Betelgeuse is also referred to as Alpha Orionis, or the brightest star in the constellation of Orion. Flamsteed's later system corrected for the limited number of letters in the Greek alphabet (twenty-four) by using a number in front of the constellation name. From Flamsteed's catalog, we have star names like 51 Pegasi, the star with the first known exoplanet, and the fifty-first brightest star in the constellation Pegasus.

Other stars possess much less poetic names. The star HD 209458 also has a planet orbiting it. The star's name begins with "HD" because it is listed in a more modern catalog, the Henry Draper catalog. Henry Draper was a nineteenth century doctor and astronomer and one of the first to use astrophotography. His widow, Anna Mary Palmer, donated funds to the Harvard College Observatory to make this catalog, which includes stars too dim for the human eye. That's why modern catalogs are much larger (and need numerical names).

When you go out and gaze at the night sky and point out the star Betelgeuse or Aldebaran, know that you're participating in thousands of years of human effort to carefully notice, name, and catalog the stars.

NOTES

Introduction

1. Thich Nhat Hanh, *The Miracle of Mindfulness: An Introduction to the Practice of Meditation* (Boston: Beacon Press, 1987), 30.
2. Mark Williams and Danny Penman, *Mindfulness: An Eight-Week Plan for Finding Peace in a Frantic World* (New York: Rodale, 2011), 229–233.
3. Jon Kabat-Zinn, *Meditation Is Not What You Think: Mindfulness and Why It Is So Important* (Boston: Hachette, 2018), xxxiv, 24.
4. Thich Nhat Hanh, *The Miracle of Mindfulness*, 3–4.

I. Getting Oriented

1. Kabat-Zinn, *Meditation Is Not What You Think*, 132.
2. Christie Taylor, "Relearning the Star Stories of Indigenous Peoples," *Science Friday*, NPR, September 6, 2019, https://www.sciencefriday.com/articles/indigenous-peoples-astronomy.
3. Aparna Venkatesan, interview with Sarah Scoles, June 1, 2020.
4. Aparna Venkatesan et. al., "Toward Inclusive Practices with Indigenous Knowledge," *Nature Astronomy* 3 (2019): 1035–1037, accessed May 30, 2020, https://www.nature.com/articles/s41550–019–0953–2.

II. The Sun

1. Thebe Medupe, interview with Sarah Scoles, June 9, 2020.
2. "Cosmic Africa Explores Africa's Astronomy," *Science in Africa*, accessed May 21, 2021, http://www.scienceinafrica.com/astronomy/cosmic-africa-explores-africas-astronomy.
3. Ferdinand Reus, "Dogon Dwellings Along the Bandiagara Escarpment," image on Wikipedia, posted March 12, 2006, https://en.wikipedia.org/wiki/Dogon_people#/media/File:Bandiagara_escarpment_2.jpg.
4. Polly W. Wiessner, "Embers of Society: Firelight Talk Among the Ju/'hoansi Bushmen," *PNAS* 111, no. 39 (September 2014): 14027–35, https://www.pnas.org/content/111/39/14027.

5. Christian Vannier, "African Cultural Astronomy," *Anthropology News*, July 12, 2019, https://www.anthropology-news.org/articles/african-cultural-astronomy/.
6. Thebe Medupe, interview with Sarah Scoles, June 9, 2020.

III. The Moon

1. Ragbir Bhathal, "Astronomy in Aboriginal Culture," *Astronomy & Geophysics* 47, no. 5 (October 2006): 5.27–5.30, https://academic.oup.com/astrogeo/article/47/5/5.27/231805.
2. James Dawson, *Australian Aborigines: The Languages and Customs of Several Tribes in the Western District of Victoria, Australia* (Sydney, Australia: George Robertson, 1881), quoted in Bhathal, "Astronomy in Aboriginal Culture," 5.27–5.30.
3. Duane Hamacher, "Lecture and Interview Highlights: Australian Indigenous Astronomy: 65,000+ Years of Science," interview by Scott Reddiex and Catriona Nguyen-Robertson, video, 10:16 (lecture, Monash Indigenous Studies Centre, Monash University, Melbourne, Australia, February 22, 2018), https://rsv.org.au/science-and-orality/.
4. Hamacher, "Lecture and Interview Highlights," https://rsv.org.au/science-and-orality/.
5. Anna Salleh, "Karlie Noon: Reaching for the Stars," ABC News, October 21, 2016, https://www.abc.net.au/news/science/2016–10–22/reaching-for-the-stars-from-the-wrong-side-of-the-tracks/7947188?nw=0; Kelly Wong, "Astronomer Karlie Noon Embraces Indigenous Science, Education and Equality," Australia's Science Channel, November 16, 2018, https://australiascience.tv/astronomer-karlie-noon-embraces-indigenous-science-education-equality/.
6. Duane Hamacher, "The Passions of Aboriginal Astronomy Guide, Willy Stevens," *Cosmos*, February 5, 2017, https://cosmosmagazine.com/space/the-passions-of-aboriginal-astronomy-guide-willy-stevens/.

IV. The Stars

1. Details for the brave: Imagine an area of the sky that is a 1 degree by 1 degree square. To cover the entire sky would take approximately 41,253 of these squares, so the entire sky has an area of 41,253 square degrees. When standing on the Earth, you can see (at best) half of this sky, so we divide that number by 2 to get 20,627. A standard toilet paper tube held up against your eye lets you see about 79 square degrees. So the number of patches it would take to cover the sky is 20,627 divided by 79, or about 250 "patches."

2. Maud W. Makemson, "Hawaiian Astronomical Concepts," *American Anthropologist* 40, no. 3 (July–September 1938): 370–83, https://www.jstor.org/stable/662037?seq=1.
3. Kālepa Baybayan, interview with Sarah Scoles, April 15, 2020.
4. "Crew Profile: Kalepa Baybayan," Polynesian Voyaging Society, accessed April 3, 2020, http://www.hokulea.com/crewmember/kalepa-baybayan/.

V. The Planets

1. Kurt Vonnegut, *A Man Without a Country* (New York: Random House, 2007), 132.
2. *The Mystery of Chaco Canyon,* directed by Anna Sofaer (Santa Fe, NM: The Solstice Project, 1999), https://solsticeproject.org/Chaco_Films_Videos/The_Mystery_of_Chaco_Canyon.
3. "About Chaco Canyon," Exploratorium, accessed May 21, 2021, https://www.exploratorium.edu/chaco/HTML/canyon.html.
4. The purpose of the National Geodetic Survey, as stated on their website, is to maintain and provide access to the National Spatial Reference System, which in turn "provides a consistent coordinate system that defines latitude, longitude, height, scale, gravity, and orientation throughout the United States and its territories." "What We Do," National Geodetic Survey website, accessed June 1, 2021, https://www.ngs.noaa.gov/INFO/WhatWeDo.shtml.
5. *The Mystery of Chaco Canyon.*
6. *The Mystery of Chaco Canyon.*
7. Willy Barreno and Isabel Hawkins, interview with Sarah Scoles, July 14, 2020.

Appendix D: Star Brightness and Color

1. Duane Hamacher, "Observations of Red-Giant Variable Stars by Aboriginal Australians," *Australian Journal of Anthropology* 29, no. 1 (April 2018): 89–107, https://onlinelibrary.wiley.com/doi/abs/10.1111/taja.12257.

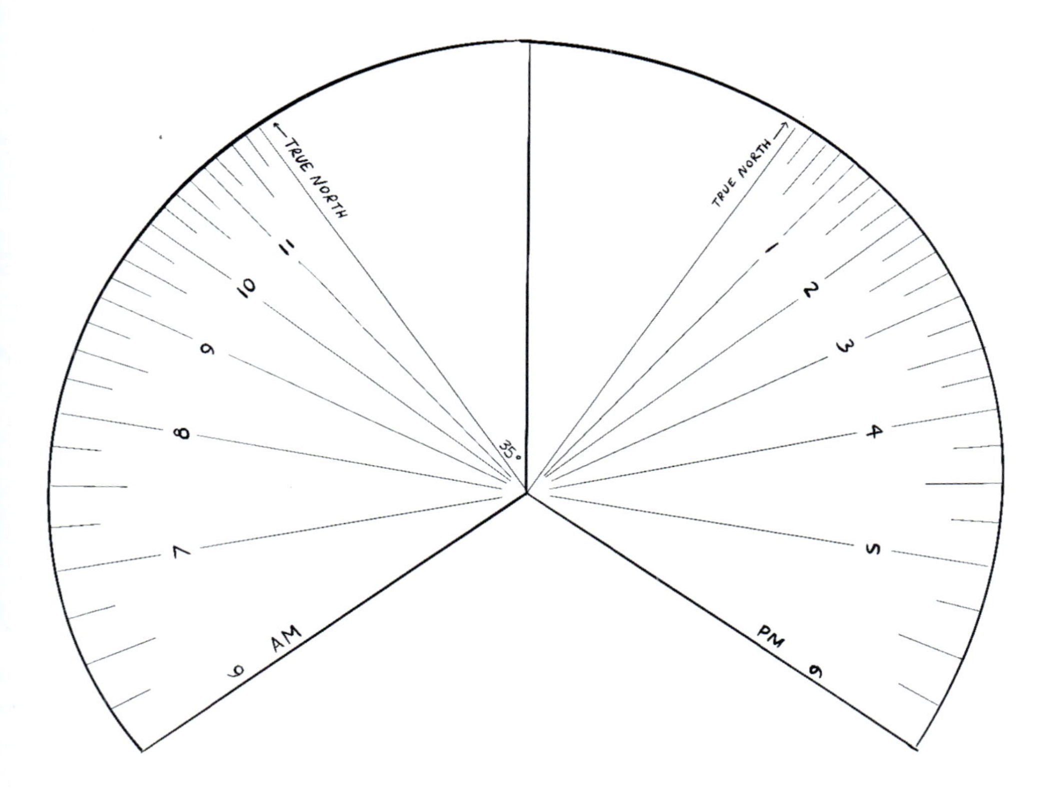

TRUE NORTH
TRUE NORTH
35°
11
10
9
8
7
6
AM
1
2
3
4
5
PM
6

Cassiopeia by NikitaRoytman Photography/Shutterstock, Inc.

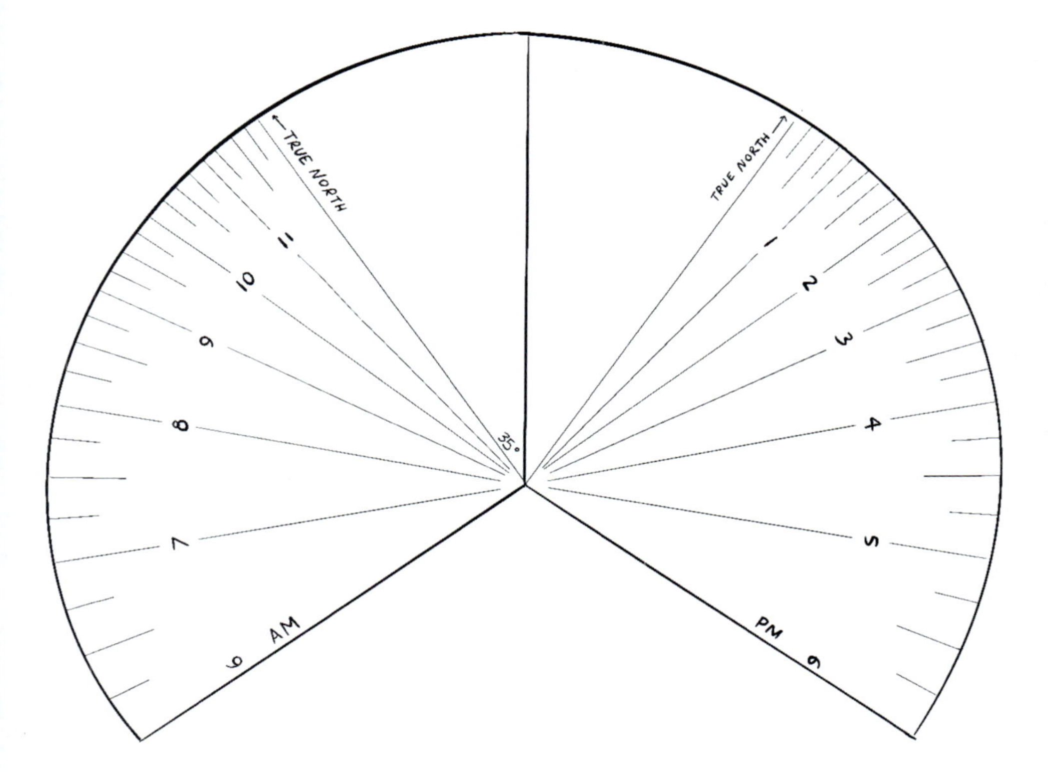

TRUE NORTH
TRUE NORTH
35°
11
10
9
8
7
6
AM
1
2
3
4
5
6
PM

Cassiopeia by NikitaRoytman Photography/Shutterstock, Inc.

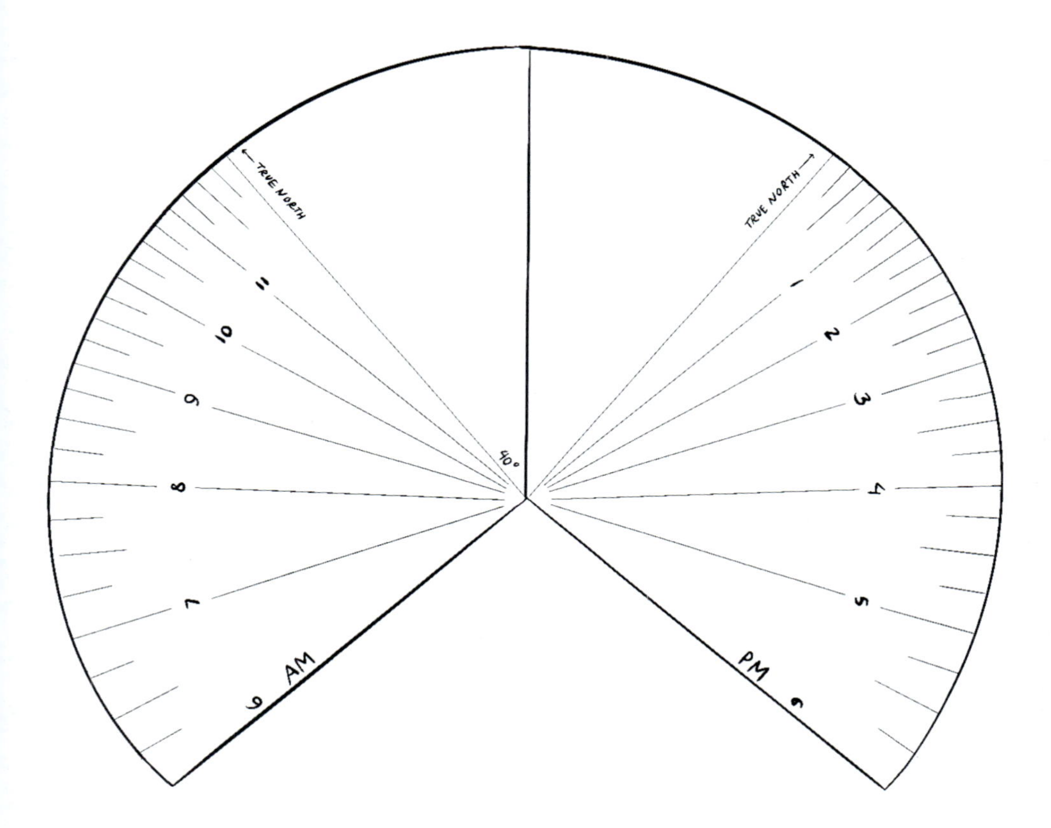
TRUE NORTH
TRUE NORTH
40°
6
AM
7
8
9
10
11
1
2
3
4
5
PM
6

Cassiopeia by NikitaRoytman Photography/Shutterstock, Inc.

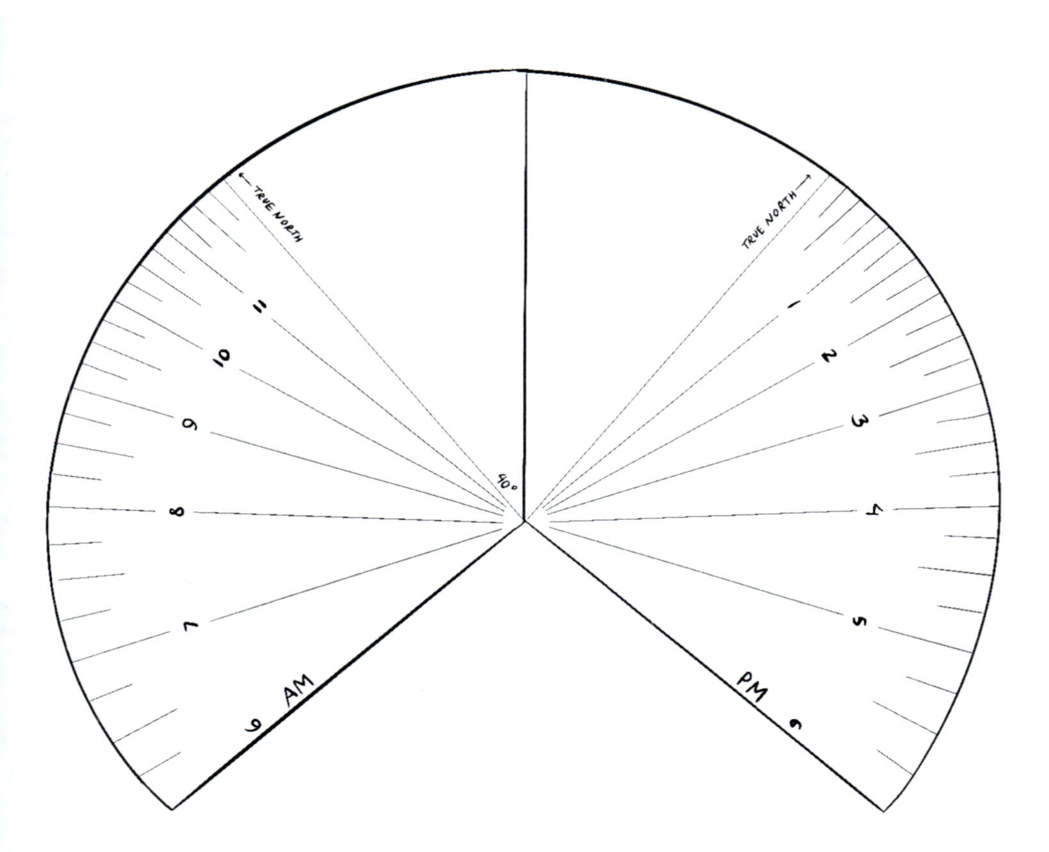

TRUE NORTH
TRUE NORTH
40°
AM
PM
6
7
8
9
10
11
1
2
3
4
5
6

Cassiopeia by NikitaRoytman Photography/Shutterstock, Inc.

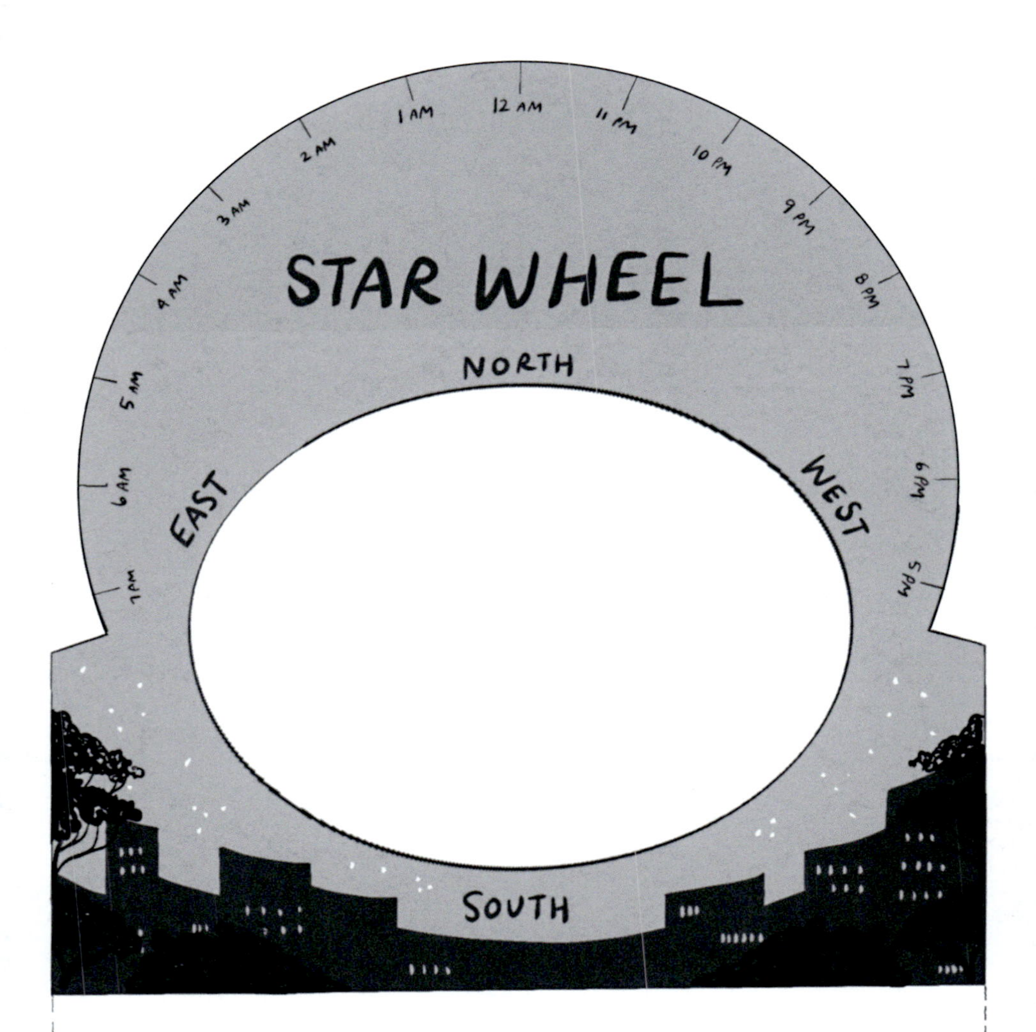

HOW TO USE:

1. ALIGN TODAY'S DATE AND TIME ON THE STAR WHEEL.
2. ROTATE YOUR STAR WHEEL SO THAT THE HORIZON YOU ARE LOOKING AT IS AT THE BOTTOM OF YOUR STAR WHEEL.
3. THE STAR POSITIONS IN THE SKY SHOULD MATCH THOSE ON THE WHEEL.
4. NOW YOU SHOULD BE ABLE TO LOCATE THE CONSTELLATIONS AND STARS IN THE SKY ABOVE.

(B)

Cassiopeia by NikitaRoytman Photography/Shutterstock, Inc.

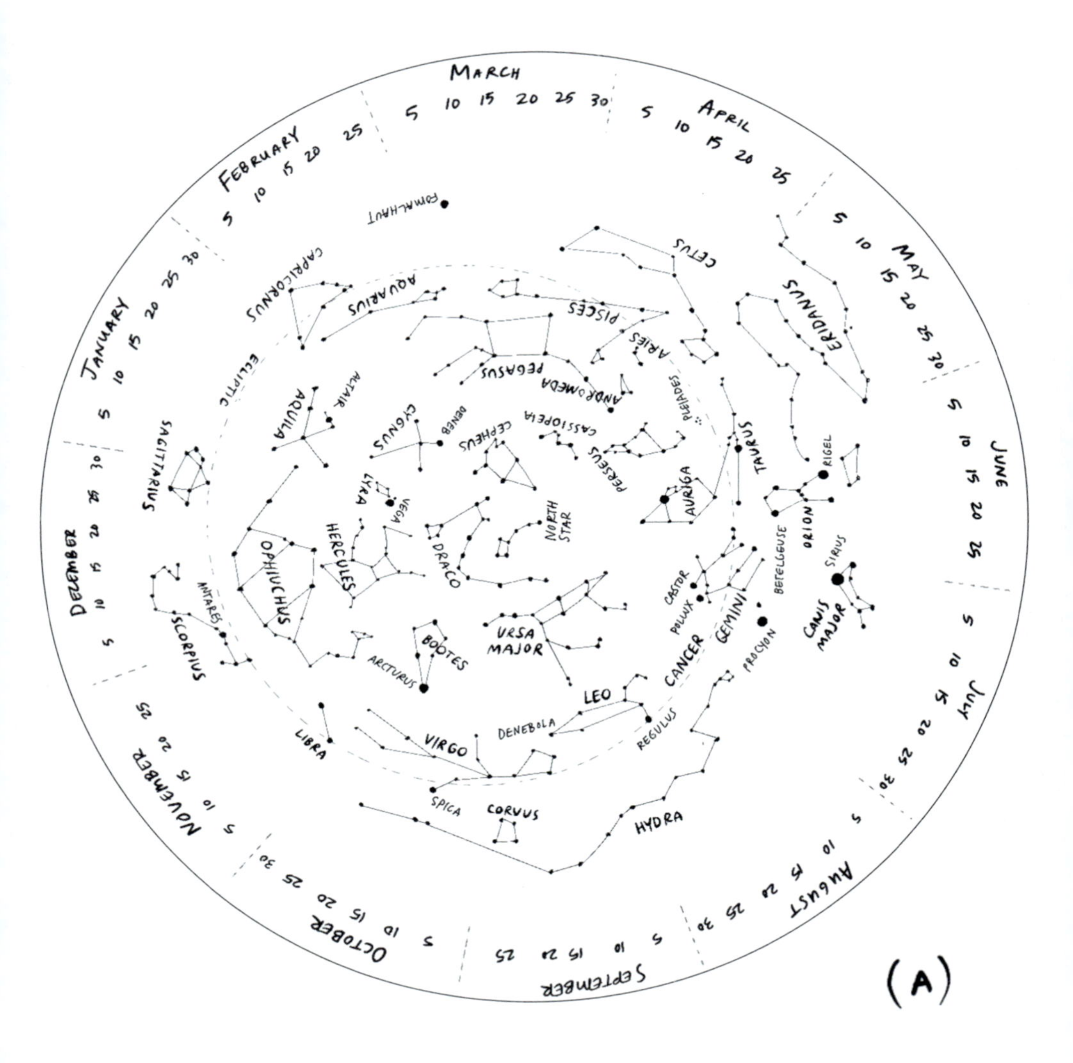

MARCH
APRIL
MAY
JUNE
JULY
AUGUST
SEPTEMBER
OCTOBER
NOVEMBER
DEZEMBER
JANUARY
FEBRUARY
FOMALHAUT
CETUS
PISCES
ARIES
ERIDANUS
CAPRICORNUS
AQUARIUS
PEGASUS
ANDROMEDA
PLEIADES
ECLIPTIC
AQUILA
ALTAIR
CYGNUS
DENEB
CEPHEUS
CASSIOPEIA
PERSEUS
TAURUS
RIGEL
SAGITTARIUS
LYRA
VEGA
AURIGA
ORION
NORTH STAR
DRACO
HERCULES
OPHIUCHUS
BETELGEUSE
SIRIUS
CASTOR
POLLUX
GEMINI
CANIS MAJOR
ANTARES
SCORPIUS
BOOTES
ARCTURUS
URSA MAJOR
CANCER
PROCYON
LEO
REGULUS
DENEBOLA
LIBRA
VIRGO
SPICA
CORVUS
HYDRA

(A)

Cassiopeia by NikitaRoytman Photography/Shutterstock, Inc.

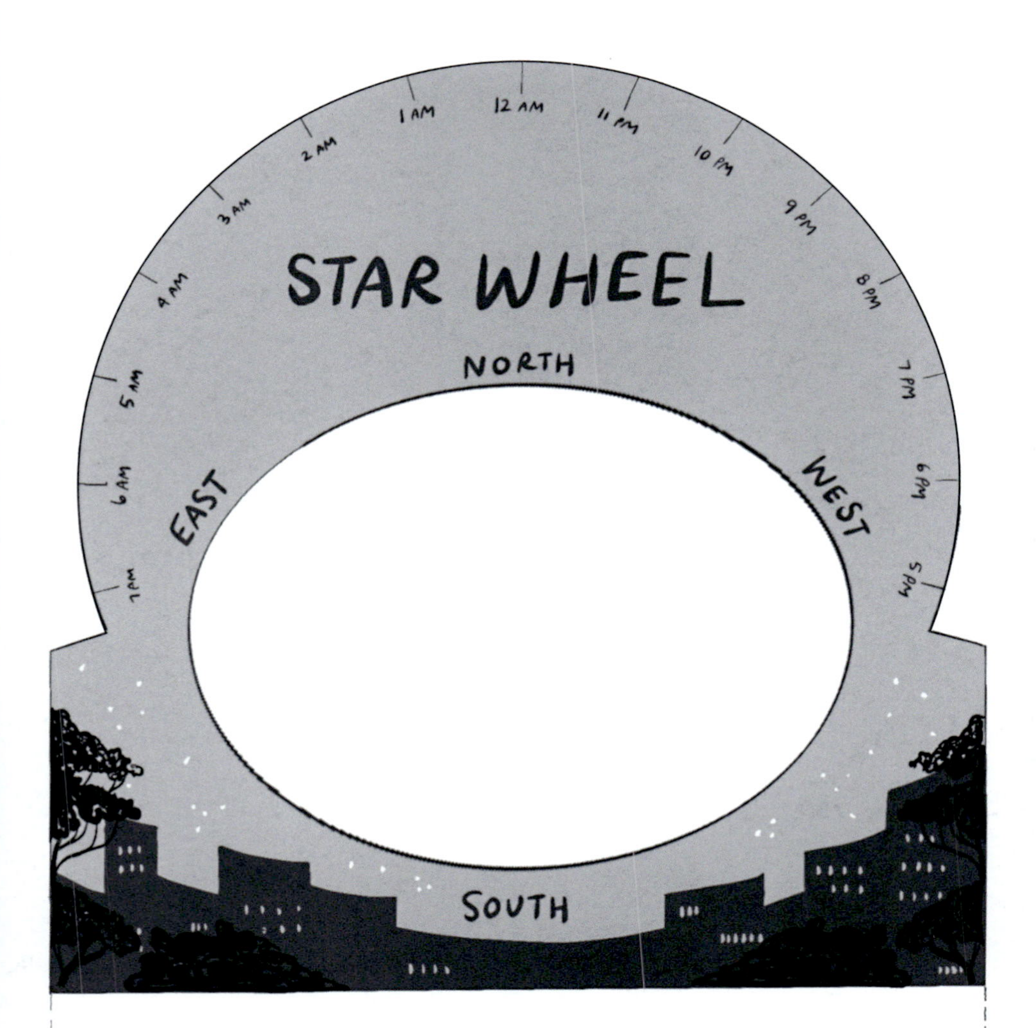

HOW TO USE:

1. ALIGN TODAY'S DATE AND TIME ON THE STAR WHEEL.
2. ROTATE YOUR STAR WHEEL SO THAT THE HORIZON YOU ARE LOOKING AT IS AT THE BOTTOM OF YOUR STAR WHEEL.
3. THE STAR POSITIONS IN THE SKY SHOULD MATCH THOSE ON THE WHEEL.
4. NOW YOU SHOULD BE ABLE TO LOCATE THE CONSTELLATIONS AND STARS IN THE SKY ABOVE.

(B)

Cassiopeia by NikitaRoytman Photography/Shutterstock, Inc.

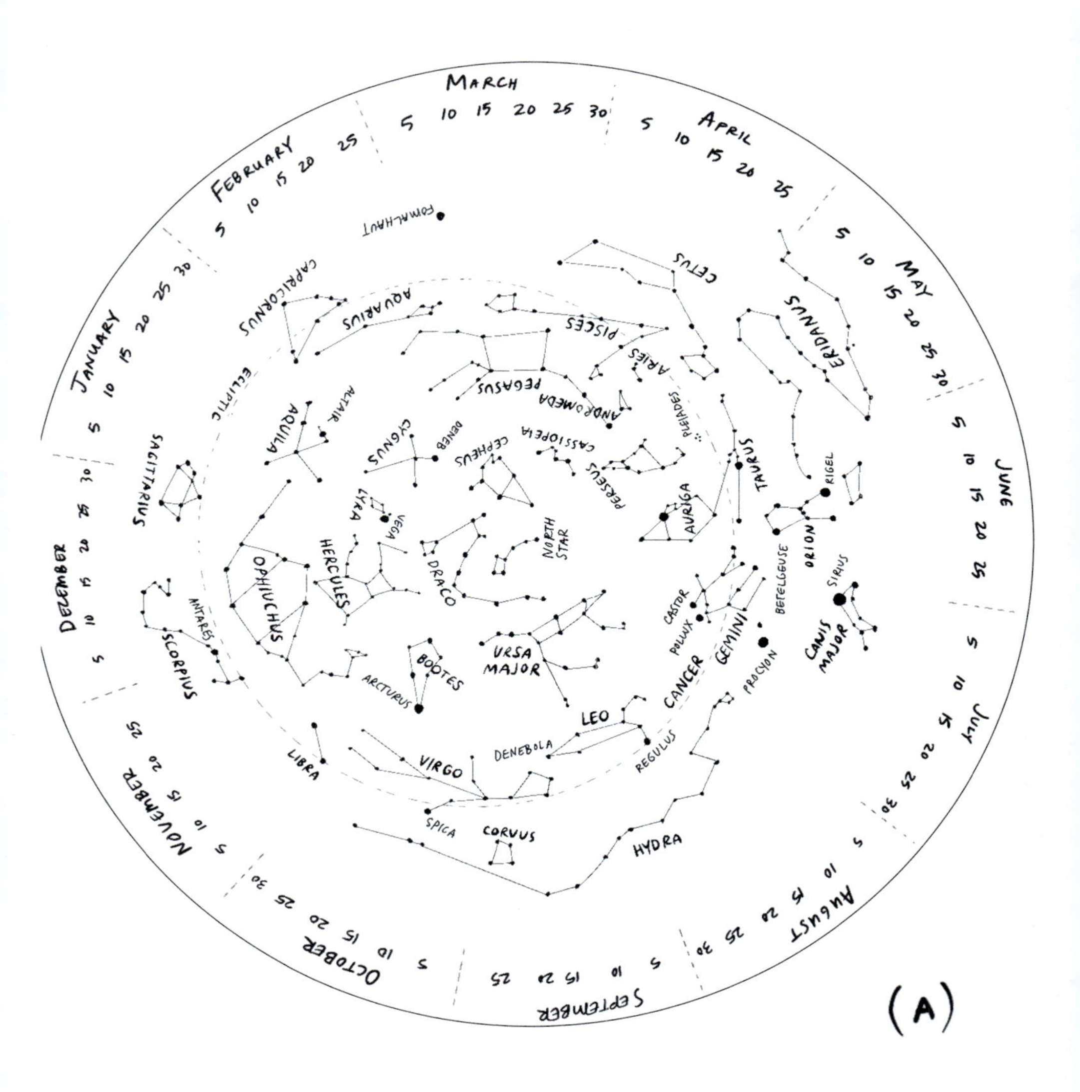

MARCH
APRIL
MAY
JUNE
JULY
AUGUST
SEPTEMBER
OCTOBER
NOVEMBER
DEZEMBER
JANUARY
FEBRUARY
FOMALHAUT
CETUS
ERIDANUS
CAPRICORNUS
AQUARIUS
PISCES
ARIES
PEGASUS
ANDROMEDA
ECLIPTIC
ALTAIR
AQUILA
CYGNUS
DENEB
CASSIOPEIA
CEPHEUS
PLEIADES
PERSEUS
TAURUS
RIGEL
SAGITTARIUS
LYRA
VEGA
AURIGA
ORION
NORTH STAR
HERCULES
DRACO
OPHIUCHUS
BETELGEUSE
SIRIUS
CASTOR
POLLUX
GEMINI
CANIS MAJOR
ANTARES
SCORPIUS
BOOTES
URSA MAJOR
CANCER
PROCYON
ARCTURUS
LEO
DENEBOLA
REGULUS
LIBRA
VIRGO
SPICA
CORVUS
HYDRA
(A)

Cassiopeia by NikitaRoytman Photography/Shutterstock, Inc.

Pleiades

VENUS

Aries

< WAXING CRESCENT MOON

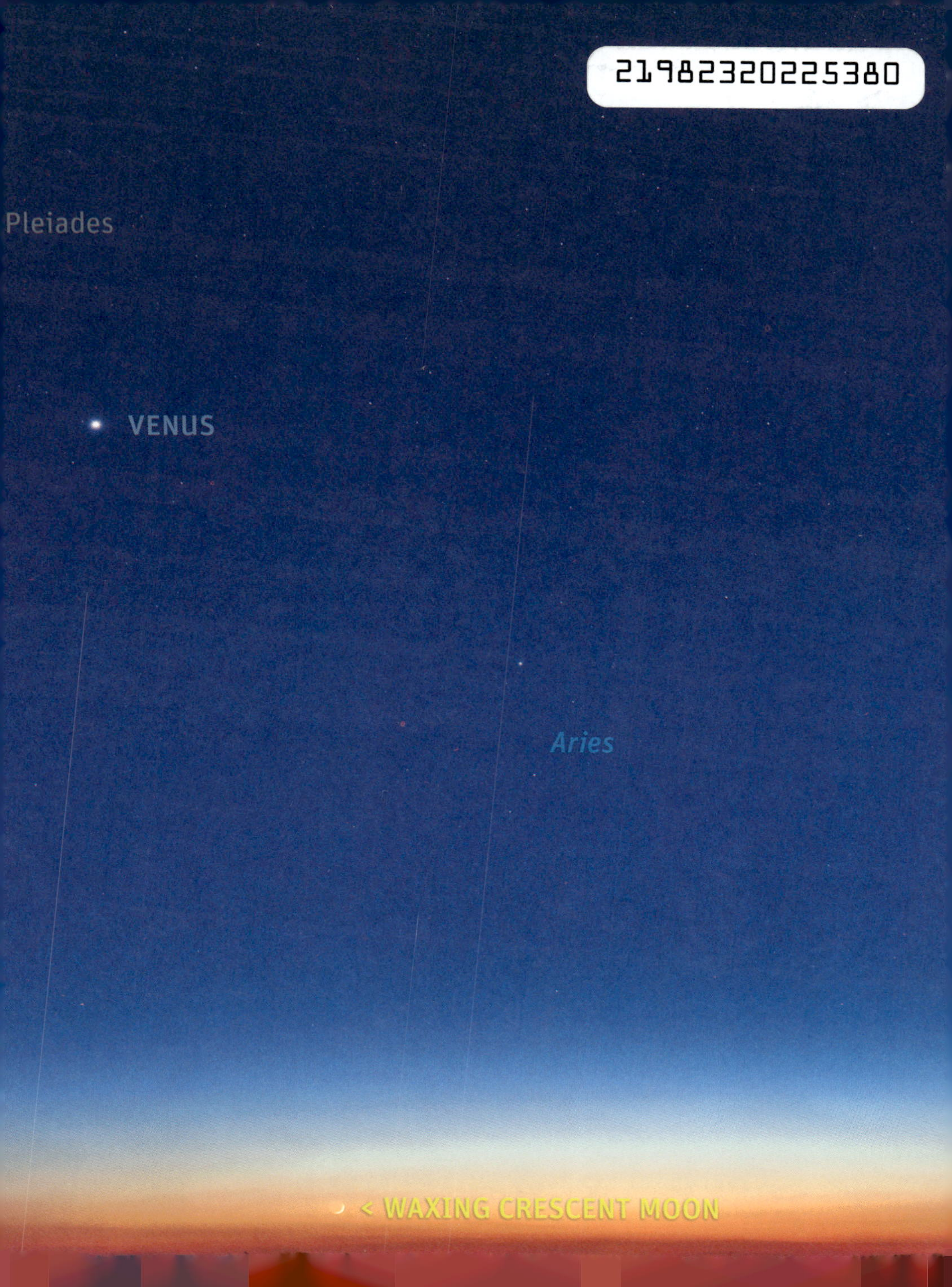